T0292348

LONDON MATHEMATICAL SOCIETY LECTURE NOTE SERIES

Managing Editor: Professor I.M. James,
Mathematical Institute, 24-29 St Giles,Oxford

1. General cohomology theory and K-theory, P.HILTON
4. Algebraic topology, J.F.ADAMS
5. Commutative algebra, J.T.KNIGHT
8. Integration and harmonic analysis on compact groups, R.E.EDWARDS
9. Elliptic functions and elliptic curves, P.DU VAL
10. Numerical ranges II, F.F.BONSALL & J.DUNCAN
11. New developments in topology, G.SEGAL (ed.)
12. Symposium on complex analysis, Canterbury, 1973, J.CLUNIE & W.K.HAYMAN (eds.)
13. Combinatorics: Proceedings of the British Combinatorial Conference 1973, T.P.McDONOUGH & V.C.MAVRON (eds.)
15. An introduction to topological groups, P.J.HIGGINS
16. Topics in finite groups, T.M.GAGEN
17. Differential germs and catastrophes, Th.BROCKER & L.LANDER
18. A geometric approach to homology theory, S.BUONCRISTIANO, C.P. ROURKE & B.J.SANDERSON
20. Sheaf theory, B.R.TENNISON
21. Automatic continuity of linear operators, A.M.SINCLAIR
23. Parallelisms of complete designs, P.J.CAMERON
24. The topology of Stiefel manifolds, I.M.JAMES
25. Lie groups and compact groups, J.F.PRICE
26. Transformation groups: Proceedings of the conference in the University of Newcastle-upon-Tyne, August 1976, C.KOSNIOWSKI
27. Skew field constructions, P.M.COHN
28. Brownian motion, Hardy spaces and bounded mean oscillations, K.E.PETERSEN
29. Pontryagin duality and the structure of locally compact Abelian groups, S.A.MORRIS
30. Interaction models, N.L.BIGGS
31. Continuous crossed products and type III von Neumann algebras, A.VAN DAELE
32. Uniform algebras and Jensen measures, T.W.GAMELIN
33. Permutation groups and combinatorial structures, N.L.BIGGS & A.T.WHITE
34. Representation theory of Lie groups, M.F. ATIYAH et al.
35. Trace ideals and their applications, B.SIMON
36. Homological group theory, C.T.C.WALL (ed.)
37. Partially ordered rings and semi-algebraic geometry, G.W.BRUMFIEL
38. Surveys in combinatorics, B.BOLLOBAS (ed.)
39. Affine sets and affine groups, D.G.NORTHCOTT
40. Introduction to Hp spaces, P.J.KOOSIS
41. Theory and applications of Hopf bifurcation, B.D.HASSARD, N.D.KAZARINOFF & Y-H.WAN
42. Topics in the theory of group presentations, D.L.JOHNSON
43. Graphs, codes and designs, P.J.CAMERON & J.H.VAN LINT
44. Z/2-homotopy theory, M.C.CRABB
45. Recursion theory: its generalisations and applications, F.R.DRAKE & S.S.WAINER (eds.)
46. p-adic analysis: a short course on recent work, N.KOBLITZ
47. Coding the Universe, A.BELLER, R.JENSEN & P.WELCH
48. Low-dimensional topology, R.BROWN & T.L.THICKSTUN (eds.)

49. Finite geometries and designs, P.CAMERON, J.W.P.HIRSCHFELD & D.R.HUGHES (eds.)
50. Commutator calculus and groups of homotopy classes, H.J.BAUES
51. Synthetic differential geometry, A.KOCK
52. Combinatorics, H.N.V.TEMPERLEY (ed.)
53. Singularity theory, V.I.ARNOLD
54. Markov processes and related problems of analysis, E.B.DYNKIN
55. Ordered permutation groups, A.M.W.GLASS
56. Journées arithmétiques 1980, J.V.ARMITAGE (ed.)
57. Techniques of geometric topology, R.A.FENN
58. Singularities of smooth functions and maps, J.MARTINET
59. Applicable differential geometry, M.CRAMPIN & F.A.E.PIRANI
60. Integrable systems, S.P.NOVIKOV *et al.*
61. The core model, A.DODD
62. Economics for mathematicians, J.W.S.CASSELS
63. Continuous semigroups in Banach algebras, A.M.SINCLAIR
64. Basic concepts of enriched category theory, G.M.KELLY
65. Several complex variables and complex manifolds I, M.J.FIELD
66. Several complex variables and complex manifolds II, M.J.FIELD
67. Classification problems in ergodic theory, W.PARRY & S.TUNCEL
68. Complex algebraic surfaces, A.BEAUVILLE
69. Representation theory, I.M.GELFAND *et al.*
70. Stochastic differential equations on manifolds, K.D.ELWORTHY
71. Groups - St Andrews 1981, C.M.CAMPBELL & E.F.ROBERTSON (eds.)
72. Commutative algebra: Durham 1981, R.Y.SHARP (ed.)
73. Riemann surfaces: a view towards several complex variables, A.T.HUCKLEBERRY
74. Symmetric designs: an algebraic approach, E.S.LANDER
75. New geometric splittings of classical knots (algebraic knots), L.SIEBENMANN & F.BONAHON
76. Linear differential operators, H.O.CORDES
77. Isolated singular points on complete intersections, E.J.N.LOOIJENGA
78. A primer on Riemann surfaces, A.F.BEARDON
79. Probability, statistics and analysis, J.F.C.KINGMAN & G.E.H.REUTER (eds.)
80. Introduction to the representation theory of compact and locally compact groups, A.ROBERT
81. Skew fields, P.K.DRAXL
82. Surveys in combinatorics: Invited papers for the ninth British Combinatorial Conference 1983, E.K.LLOYD (ed.)
83. Homogeneous structures on Riemannian manifolds, F.TRICERRI & L.VANHECKE
84. Finite group algebras and their modules, P.LANDROCK
85. Solitons, P.G.DRAZIN
86. Topological topics, I.M.JAMES (ed.)
87. Surveys in set theory, A.R.D.MATHIAS (ed.)
88. FPF ring theory, C.FAITH & S.PAGE
89. An F-space sampler, N.J.KALTON, N.T.PECK & J.W.ROBERTS
90. Polytopes and symmetry, S.A.ROBERTSON
91. Classgroups of group rings, M.J.TAYLOR
92. Simple artinian rings, A.H.SCHOFIELD
93. General and algebraic topology, I.M.JAMES & E.H.KRONHEIMER
94. Representations of general linear groups, G.D.JAMES

London Mathematical Society Lecture Note Series. 88

FPF Ring Theory

Faithful modules and generators of mod-R

CARL FAITH

Department of Mathematics, Rutgers, The State University, New Jersey

STANLEY PAGE

Department of Mathematics, University of British Columbia, Vancouver

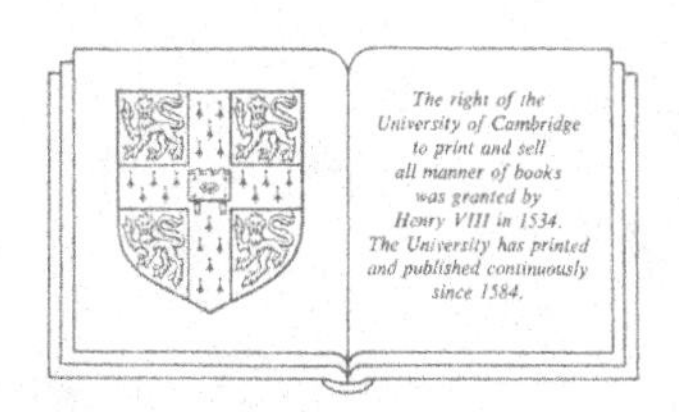

CAMBRIDGE UNIVERSITY PRESS

Cambridge

London New York New Rochelle

Melbourne Sydney

CAMBRIDGE UNIVERSITY PRESS
Cambridge, New York, Melbourne, Madrid, Cape Town,
Singapore, São Paulo, Delhi, Tokyo, Mexico City

Cambridge University Press
The Edinburgh Building, Cambridge CB2 8RU, UK

Published in the United States of America by Cambridge University Press, New York

www.cambridge.org
Information on this title: www.cambridge.org/9780521277389

First published 1984

A catalogue record for this publication is available from the British Library

Library of Congress Catalogue Card Number: 83-24067

ISBN 978-0-521-27738-9 Paperback

TABLE OF CONTENTS

PREFACE

FPF Ring Theory is the study of modules in the category mod-R of all right modules over a ring R, specifically those modules, called generators, which generate the category mod-R, and their relationship to the faithful and/or projective R-modules. Azumaya began the theory when he initiated the study of the algebras that are named after him. This led him to study generators of mod-R (called upper distinguished modules by him) and the first theorems on generators are owed to him.

Morita's seminal and monumental study of the category equivalence between mod-R and mod-S for two rings led him to many generator theorems, especially the classical Morita theorem stating that M generates mod-R iff M is finitely generated projective over its endomorphism ring $B = \text{End } M_R$, and $R \approx \text{End}_B M$ canonically (via right multiplications.)

The condition mod-R $\approx$ mod-S is called Morita Equivalence (M.E.) in his honor, and Morita's Theorem implies that this is right-left symmetric. Thus: mod-R $\approx$ mod-S iff there is a finitely generated projective generator P in mod-R, and a ring isomorphism $S \approx \text{End } P_R$. (When this is so, then $P^* = \text{Hom}_R(P,R)$ is also finitely generated projective as a canonical left R-module, and $S = \text{End}_R P^*$ canonically.)

Azumaya defined the Brauer group Br(k) over any commutative ring k. To define Br(k), consider classes of M.E. algebras, under an operation defined by

$$[A][B] = [A \otimes_R B]$$

for k-algebras A and B. This forms a semigroup S(k) and the identity [k] consists of all A such that A is M.E. to k. Now Br(k) is the group of units of S(k), and actually each $[A] \in Br(k)$ is defined by an Azumaya algebra A!

This book is mainly a study of the associative rings with the property that every finitely generated faithful module is a generator of the category of modules over the ring, called FPF rings. These rings are generalizations of pseudo-Frobenius rings (= every faithful module is a generator) which in turn are generalizations of quasi-Frobenius rings (= self-injective Artinian). This accounts for the name finitely pseudo-Frobenius (FPF). There is, moreover, a connection with the fundamental theorem of abelian groups. Namely, any ring for which each finitely generated faithful module has a free direct summand is FPF. There is, also, a finiteness condition associated with FPF rings not explicit in their name: for all known FPF rings there is a bound on the number of isomorphic one sided ideals in any direct sum contained in the ring. Rings with this property are said to be thin, and they properly include rings with finite Goldie dimension. (An infinite product of commutative self-injective rings, e.g. fields, is thin, indeed FPF, but has infinite Goldie dimension.)

A good deal of the structure of FPF rings is known but many interesting questions remain unanswered. One of the most intriguing is: are all FPF rings thin? (See Open Question for some others.)

In this volume we have organized most of the known facts concerning FPF rings and attempted to make it as self-contained as is practical.

DEDICATION AND ACKNOWLEDGEMENT

The fundamental and pioneering work of Professors Goro Azumaya and Kiiti Morita made possible this systematic study of the relationship between the concepts of "generators" and "faithful modules" of mod-R, and we dedicate this study to them.

We originally entitled this work "Azumaya-Morita Theory" until we realized how much broader than <u>FPF Ring Theory</u> that theory is.

The authors also acknowledge a great and happy debt to Professor Abraham Zaks of the Israel Institute of Technology (TECHNION) for many, many favors, both mathematical and personal. He invited both authors to Haifa, he listened to their lectures, and stimulated them with questions that exhibited his deep understanding of Azumaya-Morita theory. In particular in an early unpublished paper, Faith and Zaks proved that every commutative FPF valuation ring is quotient injective. This proved to be a proto-type theorem for commutative FPF rings. This book is based on the senior author's lecture notes "Faithful Modules and Generators of Mod-R", and he wishes to repeat his thanks given to Mrs. Marks (of Technion) for typing, and Professor John Koehl (Louisiana State University, Baton Rouge) for his critical reading, of the original manuscript. He also gratefully acknowledges Mary Ann Jablonski and "Addie" Bouillé of the Rutgers Mathematics Department staff for many favors and much help.

Many pages of the mathematics of this book were written in that wonderful coffee house in Princeton, PJ's ("A clean, well lighted place" in Hemingway's phrase.) There's no better way to thank a Herbert Tuchman (who loses money everytime one of the authors sits down!) than to tell it here. Thanks Herb, Debbie, Ruby, Alice, Joyce, Willy, Barbara, Joy, Kathy, Hilda, Karen, Liz, Patricia, Kim, Rawl, John, Tony, Maddie, and all.

The authors wish to thank Professor Page's wife Joan Marie for her patience in helping proofread this manuscript as well as his daughters for their general forebearance and finally Mary-Margaret Daisley for her superb job of typing the final manuscript.

Professor Page would like to dedicate this volume to his parents Urlin Scott and Helen Elizabeth Page, his wife Joan Marie and daughters, Marianne Elizabeth and Stephanie Theresa.

Professor Faith gives his share of the book to that talented fourteen-year-old mathematician, Japheth Wood, and to lovely Molly Sullivan.

FPF RING THEORY: FAITHFUL MODULES AND GENERATORS OF MOD-R

INTRODUCTION

In these notes we systematize the study of the property of a ring R, every (finitely generated) faithful module generates the category mod-R of all right R-modules over a ring R. Then the ring is said to be right (F)PF, or (finitely) pseudo-Frobenius.

Since R generates mod-R, and since R is a finitely generated R-module, to say that M generates mod-R is equivalent to saying that there is an onto map $M^n \to R$ of a finite direct sum of copies of M onto R. If we were to let trace(M) denote the trace ideal of M in R and M^* the dual module, then M generates R iff

$$\text{trace}_R(M) = \sum_{f \varepsilon M^*} f(M) = R$$

That is, iff there exist finitely many elements $x_i \varepsilon M$, $f_i \varepsilon M^*$, $i = 1, \cdots, n$, so that

$$\Sigma_{i=1}^n f_i(x_i) = 1$$

It is clear for a simple ring R that an R-module M generates mod-R iff trace(M) $\neq 0$, that is, $M^* \neq 0$. However, not just simple rings have this property for example: any prime ring which is Noetherian and for which nonzero ideals are invertible, has this property since, as we shall see, every nonzero right ideal is a generator of mod-R. Furthermore, since every nonzero module is faithful over a simple ring R, then right FPF implies that every simple R-module V embeds in R (via $V^* \neq 0$) and this implies that R is semisimple Artinian. Conversely, any semisimple Artinian ring is 2-sided FPF. Moreover, a Dedekind Prime Ring R is right FPF iff R is right bounded in the sense that a cyclic right module R/I is faithful only if I is an inessential right ideal. This result appears in Chapter 4; that every right FPF ring is right bounded is proved in Chapter 1.

Frobenius algebras were devised as abstractions of group algebras of finite groups over fields, but in generalizing the notion to Artinian rings, Nakayama [39] was led to the more important notion of a quasi-Frobenius ring.[1] Quasi-Frobenius rings are characterized as the Artinian rings in which annihilation defines a duality between the lattices of right and left ideals. Some twenty years later Ikeda [52] characterized quasi-Frobenius rings homologically as right self-injective right Artinian rings. (By the symmetry of quasi-Frobenius rings, they are left Artinian and left self-injective, too.) <u>Quasi-Frobenius rings are PF</u>, for over an Artinian ring the ring R embeds in a finite direct sum of copies of any faithful module, and injectivity of R splits this embedding, giving the desired epimorphism. This argument also works for any commutative self-injective ring and finitely generated faithful module, so <u>all self-injective commutative rings are FPF</u>.

Right PF rings were introduced by Azumaya in 1966 and were characterized by Azumaya, Osofsky and Utumi [66, 66, 67] as right self-injective semi-perfect rings with essential right socles. Nakayama's definition for Artinian Quasi-Frobenius rings called for a pairing of the primitive idempotents $e_i \to e_i'$ so that the top of e_iR is isomorphic to the bottom of $e_i'R$ and similarly for Re_i and Re_i', $i = 1,\ldots,n$. By a result of Faith [66] any right self-injective ring with ascending or descending chain condition on right annihilators is quasi-Frobenius, so all one sided Noetherian or Artinian PF rings are quasi-Frobenius. Osofsky [66] gave an example of a non-Artinian PF ring. Others were constructed in Faith [79a] as split-null

---

[1]Nakayama [39] refers to a paper, book, or work by Nakayama listed in References. If more than one should appear, then [39a] would refer to the first, and [39b] to the second, etc.

extensions $R = (B,E)$ of a (B,B)-bimodule E over a commutative ring B, where $(B,E) = \{\begin{pmatrix} b & x \\ 0 & b \end{pmatrix} \mid b \in B,\ x \in E\}$. Then $R = (B,E)$ is injective iff E is injective and $B = \mathrm{End}_B(E)$ canonically. Moreover, then $R = (B,E)$ is PF iff E is an injective cogenerator over B. This yields a plentiful supply of PF and FPF rings. (See Chapter 5 for these examples.)

While Osofsky's construction of a PF ring was made with the aim of showing that PF does not imply QF, nevertheless PF rings are semiperfect, which is to say semilocal rings with idempotents lifting modulo the Jacobson radical, whereas the Product Theorem for FPF Rings of Faith [79c] implies that any product of commutative, or self-basic (self-basic means semiperfect and modulo the radical a product of skew fields) FPF rings is FPF. Thus, for example, for any cardinal α, and FPF ring R, the product R^α is FPF, e.g. $\mathbb{Z}^\omega$, $\mathbb{Q}^\omega$, $\mathbb{R}^\omega$, $\mathbb{H}^\omega$, or R^ω for any field R. This provides a powerful incentive for the study of FPF rings. Not only is $\mathbb{Z}^\omega$ FPF, but R^α for <u>any</u> commutative or self-basic ring FPF ring R; for example if R is any self-basic PF ring then R^α is FPF but not PF for $\alpha \geq \aleph_0$. This subject is taken up in Chapter 1, more generally for "generic" families. (See Theorem 1.22A ff.)

Commutative FPF rings are characterized* by the two properties:

(FPF 1) Every finitely generated faithful ideal is projective.

(FPF 2) R has injective quotient ring $Q_c(R)$.

When (FPF 2) holds, R is said to be <u>quotient-injective</u>. This theorem is illustrated by the result that a domain R is FPF iff R is Prufer. (In view of the fact that a commutative self-injective ring R is FPF, the theorem indicates that finitely generated faithful ideals of a self-injective ring R are projective, but the fact is that R is the only one!)

A commutative ring R is FGC if every finitely generated module is a direct sum of cyclic modules,

---

*The results on commutative FPF rings are, for the most part, not taken up in these notes, and may be found in Faith [79a, 82a]

that is to say, the basis theorem for abelian groups extends to FGC rings.

It is easy to see that <u>any FGC ring is FPF</u>: if M is finitely generated, then $M = R/I_1 \oplus \cdots \oplus R/I_n$ for ideals $I_1 \subset I_2 \subset \cdots \subset I_n$ (in accordance with the structure theory of FGC rings [see e.g. Brandal [79] or Vamos [77]]), so M is faithful only if $I_1 = 0$, so $M \approx R \oplus X$ in mod-R as required.

Trivially, any factor ring of an FGC ring is FGC, so <u>any factor ring of an FGC ring is FPF</u>. We call the latter property CFPF, that is, a ring R is right CFPF if every factor ring is right FPF.

A ring R is said to be <u>linearly compact</u> provided that all systems

$$x \equiv x_i \bmod I_i$$

of congruences defined by ideals $\{I_i\}_{i\epsilon A}$ and elements $\{x_i\}_{i\epsilon A}$ of R are solvable if every finite subsystem is. A valuation ring R (= a chain ring, or a ring with linearly ordered ideal lattice) is said to be <u>maximal</u> provided that R is l.c., and <u>almost maximal</u> if R/I is l.c. for all ideals $I \neq 0$.

Now Kaplansky [49] proved that all almost maximal valuation rings (AMVR's) are FGC, and later raised the problem of constructing all FGC rings. The solution to the problem appears in Brandal [79], Vamos [77], and Wiegands [77]. Kaplansky [42] constructed rings of formal power series $\Sigma_{\gamma\epsilon\Gamma}\alpha_\gamma x^\gamma$ in a variable x, with coefficients α_γ in a field, and exponents γ coming from a totally ordered group Γ, and showed these rings are MVR's, i.e. there exist MVR's with arbitrary value group Γ. Thus the MVR's form an important class of CFPF rings; in fact, a local ring R is CFPF iff R is an AMVR (see, e.g. [Faith 79a]).

In 1969, Tachikawa [69] proved that a left perfect right FPF ring R is right PF, hence any right or left Artinian right FPF ring is QF. This inspired the following result of Faith [77]: if R is a semiperfect right FPF ring with nil radical, then R is right self-injective. Then Tachikawa's theorem follows from the Azumaya-Osofsky-Utumi theorem. In Chapter 2 we study semi-perfect right FPF rings. If R is semiperfect right FPF, then $R = \Sigma_{i=1}^{n} e_i R$, where $e_i^2 = e_i$, $1 = \Sigma_{i=1}^{n} e_i$ and $e_i R$ is a uniform right ideal, $i = 1,\ldots,n$. Moreover, the basic ring R_0 of R is strongly right bounded in the sense that every nonzero right right ideal contains a nonzero ideal. More can be said if the radical J of R is nil, for then R is right self-injective. (A partial converse proved in Chapter 2: Any right self-injective semiperfect ring R with strongly right bounded R_0 is necessarily right FPF.) Also, if the left zero divisors are right zero divisors for a right FPF semiperfect ring it is shown that the maximal right quotient ring is right injective and is the left classical quotient ring. This covers all known semiperfect right FPF rings.

To continue: Semiprime semiperfect right FPF rings are semihereditary and finite products of full matrix rings of finite rank over right bounded local Ore domains which are right and left valuation rings. It is shown that the basic ring of a semiperfect right CFPF ring is right duo (right ideals are two sided), right σ-cyclic (finitely generated modules are direct sums of cyclics) and finite products of right valuation rings.

Morita [58] characterized the situation when there exists a functor duality D between certain subcategories of mod-R containing R and all finitely generated R-modules, and corresponding subcategories of mod-S, for some ring S. This happens iff there exists an (S,R)-bimodule U which is an injective generator both in the category mod-R of all right R-modules and the corresponding category S-mod of left S-modules. Moreover, it is required that $R = \mathrm{End}_S U$ and $S = \mathrm{End}\, U_R$. Then R is said to possess a Morita duality, and the contravariant functor $h_U = \mathrm{Hom}_R(\ ,U)$ induces the duality D, and $\mathrm{Hom}_S(\ ,U)$ induces D^{-1}. The symbol ${}_SU_R$ is then called a Morita duality context. (Consult ART, Chapter 23 for further details.*)

The connection with PF rings is this: R is right PF iff R is an injective cogenerator in mod-R. Therefore, ${}_RR_R$ is a Morita duality context iff R is right and left PF.

Camillo-Fuller [76] characterize right (F)PF by the condition that (finitely generated) faithful right R-modules are flat as modules over their endomorphism rings. (The FPF part is implicit in their paper. But see corollary 1.19.)

On the subject of generators, a theorem of Morita [58] states that any generator M of mod-R has the property that M is finitely generated projective over its endomorphism ring. Therefore by the theorem of Camillo-Fuller, flatness-over-endomorphism ring for all (finitely generated) faithful right R-modules is equivalent to this stronger property.

By a theorem of Gabriel, if C is an abelian category with enough injectives, and if S is the left adjoint to $T:C \to D$, where D is abelian, then T preserves injectives iff S is exact. Thus, M is flat over its endomorphism ring E iff

$$\mathrm{Hom}_R(M,\ ) : \text{mod-}R \rightarrowtail \text{mod-}E$$

preserves injectives. (See Gabriel [62]; also Theorem 6.29 and its corollary in ARMC. Incredibly, the latter is related to a theorem of Bourbaki-Lambek, loc. cit. p. 28.)

In 1967, Endo [67] proved that a Noetherian commutative ring R is FPF (=(FG) in Endo's terminology) iff R is a finite product of Dedekind domains and QF local rings. Moreover, Endo also determined all right FPF rings R which are projective A-orders in a semisimple K-algebra Σ, where A is a Noetherian domain with quotient field K: R is a hereditary, maximal order in Σ. Thus right FPF implies left FPF in this case.

*ART refers to Faith [76]

In chapter 2 we show all Noetherian semiperfect FPF rings are orders in Quasi Frobenius rings. Then, using the results in Chapters 3 and 4, we show in Chapter 5 they are actually products of bounded Dedekind rings and Quasi Frobenius rings. Moreover, Endo also studied the situation where every finitely generated projective faithful R-module generates mod-R. In Chapter 4 we generalize these results and study semi-prime FPF rings satisfying other related finite conditions.

A ring R is right nonsingular if the right annihilator of each nonzero element is not essential. The maximal right quotient ring of such a ring is always a right self-injective von Neumann regular ring. (A ring is von Neumann regular if every module over it is flat.) As we have seen, not every FPF ring is self-injective; but any nonsingular PF ring is, in fact, semi-simple Artinian since the singular ideal is the Jacobson radical for self-injective rings. In Chapter 3 we study the nonsingular FPF rings. It is shown that the regular right FPF rings are characterized as the self-injective regular rings of bounded index (on the index of nilpotent elements). This says that for regular rings right FPF implies left FPF, which is not true in general (see Chapter 5). This characterization enables one to show that for right nonsingular right FPF rings the right and left maximal quotient rings coincide. Since right nonsingular right FPF rings are semiprime (and conversely) one obtains the fact that a right nonsingular right Goldie FPF ring is also left Goldie.

For modules N and M over a ring we say M has N-width a if a is the largest cardinal number such that a direct sum of a copies of N embeds in M. If for M there is a finite number ℓ such that the N-width of M is less than ℓ for all $N \neq 0$ then we say M is thin, and otherwise M is thick. A ring is right thin if it is thin as a right module. Commutative self-injective rings, and, of course Goldie rings are thin, whereas full linear rings on infinite dimensional spaces are thick. All rings with thin regular maximal quotient rings are right thin, hence every

<u>nonsingular-right-FPF-ring-is-right-thin</u>. In Chapter 5 we explore this concept and show that <u>a self-injective ring is right FPF iff it is thin and its "basic ring" is strongly bounded</u>. Thin self-injective rings have a "basic ring" much as do semiperfect rings. This allows one to parallel the theory of semiperfect FPF and CFPF rings. This is taken up in Chapter 5 where we also consider FPF group rings. (In general, a finite group G does not yield an FPF group ring RG over an FPF ring R, e.g. $\mathbb{Z}G$ is never FPF.)

All known (right) FPF rings R are (right) thin, and right quotient-injective in the sense that the classical right quotient ring $Q = Q_c^r(R)$ exists and is right injective. The problem of determining whether all right FPF rings are thin and right quotient-injective and other problems related to the structure of FPF, is appended at the end of the text.

A number of results stated above hold in the context of right $(F)P^2F$ rings, or rings over which every (finitely presented) faithful right R-module generates mod-R. A right $C(F)P^2F$ ring is one which is $(F)P^2F$ modulo any ideal. Any valuation ring R is CFP^2F, and this property characterizes VR's among local rings, in analogy with the theorem which shows that CFPF characterizes AMVR's among VR's.

1 THE BASICS

This chapter provides a format for the statements of a number of key theorems used repeatedly in the sequel, especially theorems from noncommutative ring theory which are used in Chapters 2, 3, and 5. Naturally there is a limit to what can be fitted into such a format--boredom, if nothing else, would limit any list of needed theorems--so certainly many useful theorems are relegated to the status of ad hoc citation. What follows therefore are the basics (or what has been called the bare bones!).

Because of the frequency of the references, we will abbreviate the two main references as follows:

ARMC denotes Algebra: Rings, Modules, and Categories, I.

ART denotes Algebra II: Ring Theory.

1.1A DEFINITION AND PROPOSITION.

Let mod-R denote the category of right R-modules for a ring R. An object M of mod-R is a generator iff the equivalent conditions hold:

G1. The set-valued functor $\mathrm{Hom}_R(M, \ )$ is faithful.

G2. Given an object X of mod-R, there is an index set I and an exact sequence $M^{(I)} \to X \to 0$, where $M^{(I)}$ is the co-product (= direct sum) of I copies of M.

G3. There is a finite integer $n > 0$, an object Y of mod-R, and an isomorphism $M^n \approx R \oplus Y$.

G4. The trace ideal is the unit ideal, that is,

$$\mathrm{trace}_R M = \sum_{f \in M^*} f(M) = R, \quad \text{where} \quad M^* = \mathrm{Hom}_R(M,R).$$

1.1B DEFINITION.

A ring is said to be right FPF (FP^2F) if every finitely generated (presented) faithful right module is a generator. The ring is right CFPF (CFP^2F) if every homomorphic image is right FPF (CFP^2F). PF and CPF rings are defined similarly. see 1.7a

1.1C DEFINITION AND PROPOSITION (THE MORITA THEOREM).

Let R-mod denote the left-right symmetry of mod-R. Two rings A and B are similar, or Morita equivalent, written $A \sim B$, provided that the equivalent conditions hold:

S1. mod-A $\approx$ mod-B.

S2. There exists a finitely generated projective generator P of mod-A such that $B \approx \text{End } P_A$.

S3. A-mod $\approx$ B-mod.

In the case S2, $\text{Hom}_A(P, \ )$ induces an equivalence mod-A $\approx$ mod-B and the left adjoint $\otimes_B P$ is the inverse equivalence. (The equivalence of S1-S3 is Morita's theorem [58]. Cf., Bass [62,68] or ARMC Theorem 4.29.) Also, ideals of A correspond to ideals of B in such a way that $A/I \sim B/I'$, where I' is the ideal of B corresponding to I. (See ARMC, p. 219, 4.31.3).

1.1D THEOREM (Morita).

A right R-module M generates mod-R iff M is f.g. projective over $B = \text{End } M_R$ and $R = \text{End}_B M$ canonically.

The proof of 1.1C is a bit of linear algebra. (See for example, ARMC, p.327, Prop. 7.3).

1.2A KRULL-SCHMIDT THEOREM AND EXCHANGE LEMMA.

Let

$$(1) \qquad M_1 \oplus \cdots \oplus M_n = A \oplus B$$

be a decomposition in R-mod such that End A_R is a local ring. Then, there exists i, $1 \leq i \leq n$, and an isomorphism $M_i \approx A \oplus X$ for some $X \in$ mod-R. In particular, if M_i is an indecomposable module, $i = 1,\ldots,n$, then (1) implies that $A \approx M_i$ for some i.

Let

(2) $$M = M_1 \oplus \cdots \oplus M_n = N_1 \oplus \cdots \oplus N_m$$

be two decompositions of a module M into direct sums of modules M_i and N_j each with local endomorphism rings, for each i and j. Then, $m = n$, and there is an automorphism f of M and a permutation p on n symbols such that $f(M_i) = N_{p(i)}$ $i = 1,\ldots,n$.

Furthermore, if $M = A \oplus B$, then A can be decomposed into a direct sum of modules each with local endormorphism ring.

Refer to Bass [68], or ART, pp. 39-40.

SEMIPERFECT RINGS

Let $R = \bigoplus_{i=1}^{n} e_i R$ be a direct sum decomposition of R into principal indecomposable right ideals $e_1R,\ldots,e_nR$, where e_iRe_i is a local ring, $i = 1,\ldots,n$. By definition, then, e_i is an idempotent $\neq 0$, and e_iR is an indecomposable right ideal, which we call a right prindec, for short, $i = 1,\ldots,n$. By a theorem of Bass [60], a ring R has such a decomposition if (and only if) R is semiperfect in the sense that $R/\text{rad } R$ is semi-simple, or, as we say, R is semilocal, and idempotents of $R/\text{rad } R$ lift. (Consult Chapters 18 and 20 of ART.)

THE BASIC MODULE AND BASIC RING

Now assume the notation above. Renumber idempotents if necessary so that $e_1R/e_1J,\ldots,e_mR/e_mJ$ constitute the isomorphism classes of simple right R-modules. Thus, every simple module is isomorphic to some e_iR/e_iJ with $i \leq m$ and $e_iR/e_iJ \approx e_kR/e_kJ$ iff $i = k$, for all i and $k \leq m$. The right ideal $B = e_1R + \cdots + e_mR$ is called the basic (right module of R, $e_0 = e_1 + \cdots + e_m$ is then called the basic idempotent, and $e_0Re_0 \approx \text{End } B_R$ is the basic ring of R. The basic module is unique up to isomorphism, and if f_0 is any other basic idempotent, there is a unit x of R such that $f_0 = xe_0x^{-1}$.

A projective module P is a generator iff every simple right R-module is an epic image of P (ARMC, p.148). Thus the basic module B of R is a finitely generated projective generator of mod-R, and hence, by the Morita theorem R is similar to its basic ring $R_0 = \text{End } B_R$.

A semiperfect ring R is selfbasic if R = B, in which case $R = R_0$. This condition is right-left symmetric, inasmuch as R is selfbasic if R/rad R is a product of division rings. The basic ring of R is selfbasic. The basic ring R_0 is also right-left symmetric, that is, the left basic ring $\approx R_0$.

The basic ring is a finite product of local rings iff R is a finite product of full matrix rings over local rings. In this case, R_0 is said to be local-decomposable, and R is said to be primary-decomposable. (see, for example, ART, pp.44-50).

1.2B THEOREM.

Let R be a semiperfect ring with basic right module B, and basic ring R_0. Then, R is similar to R_0. A module M generates mod-R iff B is isomorphic to a direct summand of M. Thus, if R is selfbasic[1], then M generates mod-R iff $M \approx R \oplus Y$ in mod-R.

Proof. As stated above, $R \sim R_0$ and B is a generator of mod-R, and hence so is any module containing B as a direct summand. Conversely, by Theorem 1.1A, a right module M generates mod-R iff R is isomorphic to a direct summand of M^n, for some n, and since B is a direct summand of R, we must have $M^n \approx B \oplus Y$ in mod-R. However since B is a direct sum of indecomposable modules e_iR with local endomorphism rings e_iRe_i, and $e_iR \not\approx e_jR$, $i \neq j = 1,\cdots,m$, then by the Exchange Lemma 1.2A, each e_iR

1. For the case when R has finite Goldie dimension, see Corollary 1.12B.

is isomorphic to a direct summand of M, and by repeated application of the lemma, B is isomorphic to a direct summand of M.

1.2C COROLLARY.

Under the same assumptions, an epic image M/I *of a module* M *generates* mod-R *iff* $M = B' \oplus C$ *such that* $I \subset C$ *and* $B' \approx B$. Then, $M/I \approx B \oplus C/I$.

An epic image B/I *of* B *generates* mod-R *iff* $I = 0$. *Thus, if* R *is selfbasic, then a cyclic module* R/I *generates* mod-R *iff* $I = 0$.

Proof. By 1.2B, $M/I \approx B \oplus X$ in mod-R. This means that there are submodules A and C such that $M = A + C$ and $I = A \cap C$ and $A/I = B$. By projectivity of B, I splits in A, so $A = I \oplus B'$ in mod-R, and since $I \subseteq C$, then $M = B' + C$. Since $B' \cap C \subseteq B' \cap A \cap C \subseteq K \cap I = 0$, then $M = B' \oplus C$. But $B' \approx A/I \approx B$, and hence $M/I \approx B' \oplus C/I \approx B \oplus C/I$.

By the Krull-Schmidt, or unique decomposition theorem, $B = B' \oplus C$, with $B' \approx B$ only if $C = 0$, so B/I generates mod-R only if $I = 0$. The last statement is immediate.

The next theorem states that FPF (FP^2F) are *Morita invariant* properties.

1.2D THEOREM.

A ring R *is right* FPF (FP^2F) *iff every ring* A *similar to* R *is right* FPF (FP^2F).

Proof. Let S:mod-$R \approx$ mod-A be an equivalence. As stated in 1.1B, the ideals of R and A are in a correspondence $I \leftrightarrow I'$ such that $R/I \sim A/I'$, and it follows that for any R-module M, if $I = \text{ann}_R M$, then $I' = \text{ann}_A SM$; hence M is faithful iff SM is faithful. Similarly, f.p., f.g., etc. are categorical or Morita invariant properties. (See e.g. ARMC, Chapter 2, p.92). For f.p., all that is needed is that f.g. is Morita invariant, since S preserves quotients. But a module M is f.g. iff its proper subobjects form an inductive set (ARMC, p.125, 3.8). Thus, since proper submodules are Morita invariant (ARMC, p.92,

2.4(3)) then so are f.g. modules.

1.2E THEOREM.

The properties (C)FPF and (C)FP^2F are Morita invariant.

Proof. This follows from 1.2C, and the fact that under the correspondence $I \to I'$ for ideals, that $R/I \sim A/I'$.

A ring R is *right duo*, or invariant, provided that every right ideal is an ideal. This appears to be a non-trivial and useful concept, e.g. right Noetherian right VR's are right duo and every right VR with exactly one prime ideal $\neq R$ is right duo (Brungs [69] and Brungs-Torner [77], resp.). Also, in his *Theory of Rings*, Surveys of the Amer. Math. Soc. (1943) Jacobson proved that every right and left PID is duo. Duo rings are related to regular FPF rings and semiperfect FPF rings in Chapter 3, and also in Proposition 1.5 following. Related to the duo rings are the bounded rings.

BOUNDED RINGS

1.3A DEFINITIONS.

A ring is right bounded if every essential right ideal contains a non-trivial two sided ideal. It is right strongly bounded if every nonzero right ideal contains a nonzero two sided ideal and fully right bounded if for every prime ideal the factor ring is right bounded.

1.3B PROPOSITION. (Faith [76c, 77])

Any right FPF ring R is right bounded.

Proof. Let I be any essential right ideal. If R/I is faithful, then there exists an integer $n > 0$, R-module X, and an isomorphism

$$h : (R/I)^n \longrightarrow R \oplus X.$$

Let $x_1, \ldots, x_n \in R$ be such that $h([x_1+I], \ldots, [x_n+I]) = 1$, where $1 \in R \subset R \oplus X$. If $x^{-1}I = \{a \in R \mid xa \in I\}$, then we have $\ker h = \bigcap_{i=1}^{n} x_i^{-1} I = 0$. However, $x^{-1}I$ is an essential right ideal for any $x \in R$. To see this, let $Q \neq 0$ be a right ideal. Then $xQ = 0 \to Q \subset x^{-1}I \cap Q \neq 0$. On the other hand, $xQ \neq 0$ means $I \cap xQ \neq 0$, so there is an element $y = xq \neq 0$ in $I \cap xQ$, and then $0 \neq q \in x^{-1}I \subset Q$. This contradiction proves the proposition.

The next proposition will be used in several instances in the sequel.

1.3C PROPOSITION.

If $R = \prod_{i=1}^{n} R_i$ is a finite product of right bounded rings, then R is right bounded. The converse fails. However, if R is strongly right bounded, then so is each $R_i, i = 1,\ldots,n$.

Proof. If I is any essential right ideal of say R, then $R_i \cap I, i = 1,\ldots,n,$ is an essential right ideal of R_i, and hence contains an ideal $A_k \neq 0$ of $R_i, i = 1,\ldots,n,$ and hence I contains the ideal $A = A_1 +\cdots+ A_n \neq 0$.

In the opposite direction, let $R = A \times B$ be a product of an arbitrary ring A and a field B, and let I be any essential right ideal. Then $I \cap B \neq 0$, and hence $I \cap B = B$ is an ideal of R contained in I. This proves that R is right bounded, even if A is not.

If R is strongly right bounded, then any right ideal I of R_i must contain an ideal $\neq 0$, hence R_i is strongly right bounded.

1.3D NOTE

Strongly right bounded implies that any right ideal $I \neq 0$ contains an ideal A which is essential in I.

Proof. For if A is the sum of the ideals contained in I, and if $K \cap A = 0$ for some right ideal K contained in I, then $K \neq 0$ would imply that K contains an ideal $\neq 0$, contradiction, hence $A \cap K \neq 0$.

An object M of mod-R is said to be *compactly faithful* provided that R embeds in M^n, for a finite integer $n > 0$. (In general, a module M is faithful iff R embeds in a direct product of copies of M.) Thus, by 1.1A, every generator is compactly faithful. A ring R is right

Artinian iff every module M in mod-R is compactly faithful in mod-R/A, where $A = \text{ann}_R M$. (See, for example, ART, pp. 67-69).

For a set X we will let $X^{\perp}({}^{\perp}X)$ denote the right(left) annhilator of X.

1.3E PROPOSITION

A finitely generated faithful module M over a right strongly bounded ring R is compactly faithful.

Proof. Write $M = \Sigma_{i=1}^{n} b_i R$ for elements $b_1,\ldots,b_n$ in M. Now R strongly right bounded, and M faithful, means that $\bigcap_{i=1}^{n} b_i^{\perp} = 0$, and hence $a \mapsto (b_1 a,\ldots,b_n a)$ is the desired embedding of R in M^n.

1.4 COROLLARY

Any right selfinjective strongly right bounded ring R is right FPF. (Over a right selfinjective ring R, a compactly faithful module is a generator). Thus, any semiperfect right selfinjective ring R with strongly right bounded* basic ring is right FPF.

Proof. Let M be finitely generated and faithful. By 1.3, R embeds in M^n, and then injectivity of R implies that R is a summand of M^n, so M is a generator by 1.1A. The last statement follows from the fact that R is right selfinjective iff the basic ring R_0 is. Then, apply the first statement, and Theorem 1.2D.

1.4 shows that a right FPF ring need not be semiperfect since any product of right selfinjective commutative rings will be right selfinjective and duo, hence FPF. However, an infinite such product cannot be semilocal. (Also, Z is FPF but not semilocal!)

A ring R is said to be completely right selfinjective if every factor ring is right selfinjective.

---

*Any semiperfect right FPF ring has a strongly bounded basic ring (see Theorem 2.1) Any right duo, hence any strongly regular ring, is strongly right bounded. Moreover, any right nonsingular FPF ring R has a right quotient ring $Q_{c\ell}(R)$ that is Morita equivalent to a strongly regular ring.

1.5 COROLLARY

Any completely right selfinjective right duo ring R is right CFPF.

Proof. Any factor ring of a right duo ring is right duo, and every right duo ring is strongly right bounded, so 1.4 applies.

Levy [66] gave an example of a non-Noetherian commutative ring R of which all factor rings modulo nonzero ideals are selfinjective rings, and some of the factor rings are PF. The ring exhibited is the ring R of all formal power series in a variable x indexed by the family W of all well-ordered sets of nonnegative real numbers. Thus, an element r of R has the form $r = \Sigma_{i \epsilon W} a_i x^i$, with $a_i \epsilon R$, and unique $i \epsilon W$. The only nonzero ideals of R are: the principal ideals (x^b), and those ideals of the form

$$(x^{>b}) = \{x^c u | c>b \text{ , and } u \text{ a unit of } R\} .$$

Thus, if I is any nonzero ideal, then $\bar{R} = R/I$ is completely selfinjective (and non-Noetherian). [Note, however, that we are not asserting that every cyclic $\bar{R}$-module C is injective as an $\bar{R}$-module, but merely injective as an $\bar{R}/A$-module, where $A = \text{ann}_{\bar{R}} C$. Nevertheless, C is quasi-injective as an $\bar{R}$-module. If every cyclic $\bar{R}$-module were injective as an $\bar{R}$-module, then $\bar{R}$ would be a semisimple ring by Osofsky's theorem [64]. This is not the case, since $\bar{R}$ is non-Noetherian.] Osofsky [66] gave some other examples. (See Chapter 5 for some general constructions of self-injective rings and Kaplansky [42] for more general almost maximal valuations constructed as formal power series $\Sigma_{\gamma \epsilon \Gamma} a_\gamma x^\gamma$ where Γ is a totally ordered abelian group.)

SERIAL AND QF RINGS

A ring R is said to be right serial provided that R is semiperfect and the set of submodules of every right prindec is linearly ordered. A right valuation ring (VR) is a right serial local ring. A right and left serial ring is said to be serial. An Artinian serial ring has the property that every right or left module is a direct sum of cyclic modules each of which are homomorphic images of prindecs (Nakayama [40]). A ring R is said to be (right) Σ-cyclic if every (right) module decomposes into a direct sum of cyclic modules. A ring is (right) σ-cyclic if this holds for all finitely generated modules. In particular, Artinian serial rings are Σ- cyclic. (Remark: any right Σ-cyclic ring has to be right Artinian etc.; see Chapter 20 of ART for references to this.) Warfield [75] proved that Noetherian serial rings are σ-cyclic, in fact, every indecomposable cyclic is an epic image of a prindec. A primary-decomposable serial ring, or uniserial ring, is a finite product of matrix rings over local serial rings. Asano [49] characterized Artinian uniserial rings as (right and left) Artinian (right and left) principal ideal rings.

1.6 THEOREM

A ring R is Quasi-frobenius (QF) in case R has the equivalent properties:

QF(a). Every right ideal, and every left ideal, is the annihilator of a finite subset of R.

QF(b). Every right ideal, and every left ideal, is an annihilator (= annulet), and R is right or left Artinian or Noetherian.

QF(c). R is right selfinjective, and right or left Artinian or Noetherian.

For a discussion, see, e.g., Faith [66], ART, Chapt. 24 (Note the condition QF is left-right symmetric.)

Some relationships between the various rings are: A right Artinian ring is uniserial iff every factor ring is QF. An Artinian ring R is serial iff R/J^2 is QF. For these results, see Nakayama [39, 40, 41], or ART, Chaps. 24 and 25.

The ring of lower triangular matricies $T_n(R)$ over a semisimple ring R is serial, and the injective hull of the right regular module is the full $n \times n$ matrix ring R_n. Thus, $T_n(R)$ is not selfinjective, hence not QF, hence not uniserial. However, for commutative Artinian rings, the three classes coincide: QF = serial = uniserial.

PF RINGS

For the next theorem, we remark that, dual to the result stated sup 1.2B, an injective module E is a cogenerator of mod-R iff every simple right R-module embeds in E. (See, e.g. ARMC, p.148. This follows because a module M cogenerates mod-R iff M contains the injective hull of every simple right R-module.) Thus, if R is right PF in the sense below, then every simple right R-module embeds in R.

1.7A THEOREM. (Azumaya [66], Osofsky [66], and Utumi [67]).

A ring R is right PF provided that R satisfies any of the following equivalent conditions:

(1) Every faithful right R-module generates mod-R

(2) R is an injective cogenerator for mod-R

(3) R is a semilocal right selfinjective ring with (finite) essential right socle.

The next theorem is a corollary of Theorem 2.1A and is stated here for convenient referencing.

1.7B THEOREM. (Kato [68], Faith [76b])

If R is right PF, then the basic ring R_0 is [right and left] strongly bounded. (This implies that the right socle, which is finite, say $S = V_1 \oplus \cdots \oplus V_n$, for minimal right ideals, $V_i, i = 1,\ldots,n$, is not only right essential, but also left essential, and V_i is a minimal left ideal, $i = 1,\ldots,n$).

Not only does every simple right module embed in a right PF ring, but:

1.7C THEOREM. (Kato [68])

Let R be right PF. Then every simple left module embeds in R. Thus, R is left PF iff R is left self-injective. Moreover, if R is a ring which cogenerates mod-R and R-mod, then R is right and left PF.

The last part of 1.7C is expressed by saying that a 2-sided cogenerator ring is 2-sided injective.

It is unknown whether or not right PF $\Rightarrow$ left PF. (See Corollary 5.2B in this regard.)

In the next several results CQF means every factor

ring is QF. Similarly for CPF.

<u>1.8A</u> <u>COROLLARY</u>.

(1) <u>Any QF ring is right and left PF</u>.

(2) <u>Any uniserial ring is CPF</u>.

<u>Proof</u>. (1) Apply QF(c) of 1.6 to (3) of 1.7A.

(2) Any uniserial ring is CQF.

Recall that a ring R is left perfect provided that R satisfies the d.c.c. on principal right ideals (Bass [60]) ART, Chap. 22).

<u>1.8B</u> <u>THEOREM</u>. (Osofsky [66])

<u>A two-sided PF left perfect ring R is QF</u>.

<u>1.9</u> <u>THEOREM</u>. (Tachikawa [69])

<u>A left perfect right FPF ring is right PF</u>. In Chapter 2, we present a slightly easier proof of 1.9, using 2.2B and 1.7A, and using the easy-to-prove fact that a left perfect ring has nil Jacobson radical. (2.2B asserts that a semiperfect right FPF ring with nil radical is right selfinjective, and then left perfect supplies an essential right socle.)

Actually, a two sided FPF left perfect ring is QF but we don't use this. (This is proved in Faith [76b]: Using 2.2B, we get R is left selfinjective, and then, the fact that R is right PF by 1.9 enables one to apply 1.7C to obtain that every simple left module embeds in R, hence R is a left injective cogenerator, that is, left PF, so 1.8B applies.)

<u>1.10</u> <u>THEOREM</u>. (Nakayama [40])

<u>A right or left CPF ring is uniserial</u>.

This is basically Nakayama's idea. (CF. ART, p. 238.)

<u>1.11</u> <u>THEOREM</u>.

<u>A left perfect right CFPF ring R is uniserial</u>.

<u>Proof</u>. Every factor ring has the same property, hence, by 1.9, R is CPF, and therefore uniserial by 1.10.

FINITE GOLDIE OR UNIFORM DIMENSION

A submodule S of a module M is <u>essential</u> provided that

$$S \cap K = 0 \Rightarrow K = 0$$

$\forall$ submodules K. Otherwise S is said to be <u>inessential</u>. An <u>essential right ideal</u> of R is an essential submodule of the right R-module R.

A submodule K of module M is said to be a <u>complement submodule</u> provided that there is a submodule S such that K is maximal in the set of all submodules T such that $S \cap T = 0$. In this case, K is said to be a complement of S. Note that K is not unique in general. Every submodule has at least one complement by an application of Zorn's lemma.

A <u>double complement</u> of a submodule S is a complement containing S of some complement of S. Thus, if K is a complement of S, then by Zorn's lemma, there is a complement Q of K containing S. Moreover, Q is an essential extension of S. This can be used to prove that the only complement submodules of an injective module are the direct summands. (The converse holds in any module: any direct summand is a complement.)

A nonzero module M is <u>uniform</u> provided that $I \cap K \neq 0$ for any two nonzero submodules I and K, or equivalently, every nonzero submodule is an essential submodule. Thus, M is uniform iff the injective hull $\hat{M}$ is uniform iff $\hat{M}$ is indecomposable iff End $\hat{M}_R$ is a local ring. Thus, M is uniform iff 0 and M are the only complement submodules.

<u>1.12A GOLDIE DIMENSION THEOREM</u>.

<u>For a module</u> M <u>and its injective hull</u> E, <u>the f.a.e.:</u>

(1) M <u>satisfies</u> (acc) $\oplus$, <u>that is, every nonempty set of independent submodules</u> M <u>is finite</u>

(=M contains no infinite direct sum of nonzero submodules.)

(2) E is a direct sum of finite number of indecomposable modules.

(3) M contains an essential submodule which is a direct sum of a finite number of uniform submodules.

(4) M has the a.c.c. on complement submodules.

When this is so, then there is a maximal integer n in (1) - (3) (the same in all three), and n is called the Goldie dimension of M. Then, every ascending chain of complement submodules has length $\leq n$.

This follows from the Krull-Schmidt Theorem, or Unique Decomposition Theorem (see, e.g., ART, p.40; also ARMC, pp. 344-5).

1.12B COROLLARY.

Let R be a semiperfect ring with basic module B. If R has finite Goldie dimension m, then B has Goldie dimension $n \leq m$, and any generator, M of mod-R has Goldie dim $M \geq n$, with equality holding iff $M \approx B$. In particular, a submodule I of B generates mod-R iff $I \approx B$.

Proof. Since B is a direct summand of R, then $n \leq m$. By Theorem 1.2B, $M \approx B \oplus X$ in mod-R, hence dim $M \geq n$, with equality holding iff $X = 0$, that is, iff $M \approx B$. If M embeds in B, then equality must hold, hence the last statement.

NONSINGULAR MODULES AND RINGS AND MAXIMAL QUOTIENT RINGS

Much of what follows is anthologized in ARMC, Chapters 9 and 10, and in ART, Chapter 19.

The singular submodule sing M of a module M is defined by:

$$\text{sing } M = \{x \in M \mid x^{\perp} \text{ is an essential right ideal}\} = Z(m)$$

where

$$x^{\perp} = \mathrm{ann}_R x = \{a \in R \mid xa = 0\} .$$

Since sing M is a fully invariant submodule of M, then the right singular ideal of R, defined as sing R_R, is indeed an ideal of R. Note sing R $\neq$ R since 1 $\notin$ sing R_R. A module M is nonsingular if sing M = 0, and R is a right nonsingular ring if sing $R_R = 0$. If R is right nonsingular, then M/sing M is nonsingular for any M. A ring is nonsingular if both right and left nonsingular.

Let $E = \hat{R}$ denote the injective hull in mod-R of R. Then, every essential extension of R embeds in E. Now a module M is a rational extension of R provided that $\mathrm{Hom}_R(S/R,M) = 0$ for any submodule S containing R, or equivalently, for each pair $x,y \in M$, with $y \neq 0$, there corresponds $r \in R$ such that $xr \in R$, with $yr \neq 0$. (This is a specialization to R of the general notion of Findlay-Lambek). It follows that any rational extension is an essential extension, hence embeds in E, and moreover, is contained in $\bar{R} = \mathrm{ann}_E(\mathrm{ann}_B R)$, where $B = \mathrm{End}\ E_R$. Further,

$$\bar{R} = \{x \in E \mid b(x) = 0 \ \forall\ b \in B : b(1) = 0\}$$

is the maximal rational extension of R in mod-R, and $\bar{R}$ is a ring isomorphic to $\mathrm{End}_B E$ under a map f such that $h^{-1}(f) = f(1)$ for all $f \in \mathrm{End}_B E$. (See ART, Proposition 19.34, or consult Lambek [66].) This ring is called the (Johnson-Utumi) maximal right quotient ring of R contains R as a subring, and is denoted variously $\bar{R}$ or $Q^r_{max}(R)$ or simply Q^r_{max}. The full, or classical right quotient ring $Q^r_c(R)$, when it exists (sup. 1.12B) embeds in Q^r_{max}. Furthermore, when R is right nonsingular, then $E = \hat{R}$ is itself a rational extension of R, hence $\hat{R} = Q^r_{max}$ is injective. Moreover, $Q = \hat{R}$ is a von Neumann regular right selfinjective ring. Therefore a right ideal J of Q is a complement right ideal (= submodule) of Q iff $J = eQ$ with $e = e^2 \in Q$. Each right ideal I of R is contained in a unique maximal essential extension, or injective hull, $\hat{I}$ in $Q = \hat{R}$, and $\hat{I}$ is a complement right ideal of Q, hence generated by a idempotent. Then $\bar{I} = \hat{I} \cap R$ is the unique maximal essential extension of I in R; $\bar{I}$ is a complement right ideal of R, and is the least complement right ideal of R containing I. Thus, the mapping $\bar{I} \to \hat{I}$ is a lattice isomorphism between complements of R and complements of Q, (and contraction is the inverse mapping). In particular, since Q is a von Neumann regular ring, then Q has the a.c.c. on right complements iff Q has the d.c.c. on right complements iff Q is semisimple Artinian. In this case, then R has the a.c.c. and d.c.c. on complement right ideals.

ANNIHILATOR RIGHT IDEALS

An annihilator right ideal, or right annulet, I is one of the form $X^{\perp}$ for a subset X of R, and then $I = L^{\perp}$, where $L = {}^{\perp}I$ is the left annihilator of I. Thus, the mapping $I \to {}^{\perp}I$ defines a lattice isomorphism between the lattices of right and left annulets. When R is right nonsingular, then any right annulet I is a complement right ideal (Inasmuch as patently $I = eQ \cap R$, where $\hat{I} = eQ$, and e idempotent. This follows, since if $J = \{q \in Q | qI = 0\}$, then $J = Q(1-e)$, and hence $I = \operatorname{ann}_Q J \cap R = eq \cap R$.) It follows that the a.c.c. on complement right ideals of R implies the a.c.c. and d.c.c. on right complements and right annulets.

The question arises, when is $Q^r_{max}(R) = Q^{\ell}_{max}(R)$? The answer by Utumi [63] that every right or left annulet is a complement characterizes this condition. Moreover, every right complement ideal is a right annulet iff every nonzero left ideal of Q meets R. (Cf. Theorem 2.16A.)

$ACC^{\perp}$ AND (ACC) ⊕

We let $(acc)^{\perp}$ (resp. $^{\perp}(acc)$) denote the condition that R satisfies the a.c.c. on right (resp. left) annulets. The corresponding symbol for the a.c.c. on complement right (resp. left) ideals is (acc) ⊕ (resp. ⊕ (acc)), and this is equivalent to the a.c.c. on direct sums of right (left) ideals contained in R. A ring is called right Goldie if it satisfies both (acc) ⊕ and $(acc)^{\perp}$. Any right nonsingular ring R is right Goldie iff $\hat{R}$ is semisimple iff R has (acc) ⊕ . (This follows from the above stated lattice isomorphism $\bar{I} \to \hat{I}$ between the lattice of complement right ideals of R and $Q = \hat{R}$.)

CLASSICAL QUOTIENT RINGS

A ring R has a classical or full right quotient ring, denoted $Q_{c\ell}(R)$, or just $Q_{c\ell}$, if there is a ring embedding $R \to Q_{c\ell}$ such that every regular element a of R is invertible in $Q_{c\ell}$, and, moreover, $Q_{c\ell} = \{ba^{-1} | b \in R,$ regular $a \in R\}$. A necessary and sufficient condition for the existance of $Q_{c\ell}$ is the right Ore condition:

$$\forall\, b \in R \text{ \& regular } a \in R\ \exists\, b_1 \in R \text{ \& regular } a_1 \in R \text{ such that } (ba_1 = ab_1)$$

(Thus, in $Q_{c\ell}$ we must have $b^{-1}a = a_1 b_1^{-1}$.) Then R is called a right Ore ring. Clearly, any commutative ring is an Ore ring. An integral domain is a right Ore ring iff it is right Goldie iff it has (acc) ⊕ , and in this case $Q_{c\ell}$ is a field. Thus, a domain R is right Ore iff R is uniform iff R has Goldie dimension = 1 in mod-R.

1.12C GOLDIE'S THEOREM.

A ring R is a right Ore ring with semisimple (Artinian) right $Q_{c\ell}$ iff R is a semiprime right Goldie ring. In this case $Q_{c\ell}$ is simple Artinian iff R is prime. In either case, R is left Ore iff R is left Goldie, and in this case $Q_{c\ell}$ is the left quotient ring.

As stated, $Q^r_{c\ell}$ embeds in Q^r_{max}, with inclusion being strict in general. However a semiprime right Goldie ring is right nonsingular with $Q^r_{max} = Q^r_{c\ell}$. Conversely, if R is a right nonsingular right Goldie ring, then $Q^r_{max} = Q^r_{c\ell}$ iff R is semiprime. A ring R is Goldie provided that it is both right and left Goldie.

PRINCIPAL IDEAL RINGS

1.13 THEOREM. (BASIS THEOREM FOR PIR'S)
A right and left principal ideal ring is σ-cyclic.

Proof. The proof requires Goldie's theorem [62] (cf. ART, p.128, Theorem 20.37). Noetherian, principal right ideal rings, namely, $R \approx A \times B$, where A is a semiprime ring, and B is an Artinian ring. (For commutative R, this is a theorem of Asano.) For the Artinian principal ring B , another theorem of Asano [49] states that B is a finite product of primary serial rings, so the theorem of Nakayama [41] applies: every module is a direct sum of uniserial modules. B is therefore Σ-cyclic. (Kaplansky's theorem [49] also states that B is Σ-cyclic in the commutative case.)

This reduces the proof to the case $R = A$ is a semiprime PIR, but in this case, we first only require that R be a principal right ideal ring (right PIR). For then, Goldie's Theorem (loc. cit.) asserts that R is a finite product of prime right PIR's, and another of his theorems states that a prime PIR is isomorphic to a full $n \times n$ matrix ring F_n over a right Noetherian right Ore domain. (Loc. cit.; also ARMC, p.411, Theorem 10.21). Furthermore, R is then right hereditary. (For every essential right ideal I contains a regular element. If $I = aR$ contains the regular element x, then a is itself regular, so $I \approx R$ is projective. But every right ideal is a summand of an essential right ideal, so R is right hereditary.)

To continue, we must assume that the classical quotient ring is a two-sided quotient ring, that is, that R is left Ore. (By Theorem 1.12, this is the case when R is left Goldie, e.g., when R is left Noetherian). When this is so, then any finitely generated module decomposes into a direct sum

$$M \approx M/t(M) \oplus t(M)$$

where $t(M)$ is the torsion submodule, namely the set of all elements of M annihilated by regular elements. This is possible since the latter set is a submodule, and, moreover, $M/t(M)$ can be embedded in a free module, hence is projective. (See Levy [63], p.149, where this is stated for a right semihereditary right and left Goldie semiprime ring.)

By Kaplansky's theorem (Cartan-Eilenberg [56], Chapter 1), over any right hereditary ring, any projective module is isomorphic to a direct sum of right ideals, which in our case are principal, so $M/t(M)$ is σ-cyclic. Moreover, by a theorem of Eisenbud and Griffith [71] (also in ART, Chap. 25), for any right and left Noetherian hereditary prime ring R, R/I is a serial Artinian ring for any ideal $I \neq 0$. Thus, if $t(M) \neq 0$, then we can take $I = \text{ann}_R t(M) = t(M)^{\perp}$ and see that $t(M)$, whence M, is σ-cyclic.

The proof for the semiprime case of the theorem yields the corollary:

<u>1.14</u> <u>COROLLARY</u>

<u>Any left Noetherian principal right ideal semiprime ring is σ-cyclic</u>.

S. Singh [75] has provided a converse to the Eisenbud-Griffith theorem cited above. if R is a bounded Noetherian prime ring such that every factor ring is serial Artinian, then R must be hereditary.

MODULES FLAT OVER ENDORMORPHISM RING: THE THEOREM OF CAMILLO AND FULLER

Camillo and Fuller [76] characterized the condition that every (finitely generated) faithful left R-module is finitely generated and flat over its endomorphism ring by the requirement that the ring is left (F)PF. The proof uses the notion: a module U generates a module M provided that M is an epic image of a direct sum of copies of U, or equivalently,

$$M = \sum_{f \in \mathrm{Hom}_R(U,M)} f(U) .$$

Thus, by 1.1A, a module U will be a generator iff U generates every module iff U generates R.

A test for flatness (Chase [70]; also ARMC, p. 438, 11.33) states that a module U is flat over a ring Δ iff for every relation $\Sigma_{j=1}^{n} x_j\gamma_j = 0$ $(x_j \in U,\ \gamma_j \in \Delta)$ there exist $y_i \in U$, $\varepsilon_{ij} \in \Delta$ $(i=1,\ldots,m, j=1,\ldots,n)$ such that, for each $j\in\{1,\ldots,n\}$ and each $i\in\{1,\ldots,m\}$, $\Sigma_{i=1}^{m} y_i\varepsilon_{ij} = x_j$ and $\Sigma_{j=1}^{n} \varepsilon_{ij}\gamma_j = 0$. An interpretation of this test for $\Delta = \mathrm{End}_R U$ gives the following proposition of Camillo and Fuller [76].

1.15 PROPOSITION.

A left R-module U is flat over its endomorphism ring if and only if it generates the kernel of each R-homomorphism $d : U^n \to U$ $(n = 1,2,\ldots)$.

Proof. If $\Delta = \mathrm{End}_R U$, then $\mathrm{Hom}_R(U^m,U^n)$ is the set of functions

$$(u_1,\ldots,u_m) \longmapsto (u_1,\ldots,u_m)\ [\delta_{ij}]_{m\times n}$$

given by ordinary matrix multiplication. Viewing

$d = \begin{bmatrix} \gamma_1 \\ \vdots \\ \gamma_n \end{bmatrix} \in \mathrm{Hom}_R(U^n,U)$ and $[\varepsilon_{ij}]_{m\times n} \in \mathrm{Hom}_R(U^m,U^n)$ in this light, we see that the lemma is an immediate consequenc of the stated flatness test.

This also characterizes when $\mathrm{Hom}_R(U,\ )$ preserves injectivity.

1.16 COROLLARY

A left R-module U has the property that $\mathrm{Hom}_R(U,\ ) : R\text{-mod} \to \Delta\text{-mod}$ preserves injectives, where $\Delta = \mathrm{End}_R U$, iff U generates the kernel of each homomorphism $U^n \to U$.

The corollary follows from the lemma and a theorem on abelian categories with enough injectives, namely that if $T: C \to D$ has a left adjoint S, then S is exact iff T preserves injectives, assuming C has enough. (See, e.g., ARMC, p.318, 6.28).

A theorem of Popescu-Gabriel states that if C has exact direct limits and a generator U, then $\mathrm{Hom}_C(U,\ )$ has exact left adjoint. (See, ARMC, p.515 for references.)

We note that $\mathrm{Hom}_R(S,\ ) : \text{mod-}R \to \text{mod-}S$ preserves injective hulls where S is a ring, and R a subring. Consult Cartan-Eilenberg [56], II, Prop. 6.1A, Eakin [68], Eisenbud [70], Hosaka-Ishikawa [73], and Formanek-Jategoankar [74] for conditions under which $\mathrm{Hom}_R(S,\ )$ preserves injective hulls. (Also see the proof of Theorem 5.21.)

By using Morita equivalence or the similarity $\mathrm{End}_R U \sim \mathrm{End}_R U^n$ one can replace $d : U^n \to U$ by $d \in \mathrm{End}_R U^n$ in Proposition 1.15.

As before the injective envelope of a module M is denoted by $E(M)$.

1.17 COROLLARY.

A ring R is left self-injective if and only if $E(R) \oplus E(R)/R$ or $E(R) \oplus E(E(R)/R)$ is flat over its endomorphism ring.

Proof. If R is injective then $E(R) \oplus E(E(R)/R) = R$, so one implication is trivial.

Conversely, apply 1.15 to the endomorphism

$$(x,y) \to (0, x + R) \quad \text{of} \quad E(R) \oplus E(R)/R \quad \text{or} \quad E(R) \oplus E(E(R)/R)$$

to see that R is (generated by) an injective module.

An interpretation of the next result is that if the faithful R-modules provide reasonably "nice" representations for R as modules over their endomorphism rings, then R is PF.

1.18 THEOREM. (Camillo and Fuller [76]).

A ring R is left PF if and only if each of its faithful quasi-injective left modules is finitely generated flat over its endomorphism ring. (When R is left PF, then each faithful module is f.g. projective over its endomorphism ring.)

Proof. Let $C = \oplus_A E(T_\alpha)$ where $(T_\alpha)_{\alpha \varepsilon A}$ represents one copy of each simple left R-module. Let $E = E(C)$, $\Delta = \mathrm{End}_R E$ and $Q = C\Delta \leq {}_R E_\Delta$. Then Q is quasi-injective, generated by C, and faithful. By hypothesis, Q is finitely generated over its endomorphism ring B, and $\mathrm{Hom}_B(,Q)$ converts $B^n \to Q \to 0$ exact in mod-B into $0 \to R \to Q^n$ exact in R-mod. By 1.17, R is injective, so this embedding splits, and hence Q generates R. But C generates Q, so C is a generator, Since R therefore embeds in C^n for some integer $n > 0$, then R has essential left socle, so 1.7a implies that R is left PF. The converse follows from the fact that a generator U is always f.g. projective over its endomorphism ring. (See 1.10).

Similarly, one characterizes left FPF rings:

1.19 COROLLARY.

A ring R is left FPF iff every f.g. faithful left R module is f.g. flat over its endomorphism ring.

Proof. The necessity follows as in the proof of the theorem: every generator is not only flat but (f.g.) projective over its endomorphism ring. Conversely, if M is f.g. over its endomorphism ring, say $B^r \to M \to 0$ exact in mod-B, then by the argument of 1.18, R embeds in M^n. Since $U = M^n \oplus M^n/R$ is also flat over its endomorphism ring, and since there is an exact sequence

$$0 \to R \oplus M^n/R \to U \xrightarrow{f} U$$

then by Proposition 1.15, U generates the kernel, hence U generates R . But, M generates U , so M generates R , proving R is left FPF.

THE GENUS OF A MODULE
AND
GENERIC FAMILIES OF RINGS

A right module M over a ring R is said to have a <u>unimodular element</u> (UME) if there exists $u \in M$ such that uR is a direct summand of M canonically isomorphic to R. Thus, M has a UME iff there is an epimorphism $M \twoheadrightarrow R$. In general, a module M generates the category mod-R of all right R-modules iff there is an epic $M^n \twoheadrightarrow R$ for some integer $n > 0$; equivalently, M^n has a UME. In this case, we let $\gamma(M)$ denote the infimum of all such integers n, and call this the <u>genus</u> of M. If M does not generate mod-R, we set $\gamma(M) = \infty$. The <u>(little) right genus</u> of a ring R will be denoted by $g_r(R)$ and is defined to be the supremum of $\gamma(M) < \infty$ for M finitely generated in mod-R. The <u>big right genus</u> $G_r(R)$ is defined similarly without restriction on finite generation of M. Clearly, $g_r(R) \leq G_r(R)$, and equality holds when R is a right Noetherian ring. (Faith [79c])

A family $F = \{R_i\}_{i \in I}$ of rings is <u>generic of (with) bound</u> B or <u>right</u> B-<u>generic</u> if there exists a function $B : \mathbb{Z}^+ \to \mathbb{Z}^+$ such that for all modules M if $\gamma(M) < \infty$ and $\nu(M) < \infty$ is the minimal number of elements in any set of generators of M, then $\gamma(M) \leq B(\nu(M))$. The product theorem states that any product of a generic family of rings of bound B is a ring which is generic of bound B (considering a ring as a family with one member) (see Theorem 1.20).

For example, a family of rings each of genus

$\leq g$ is generic with bound $\leq g$, where g also denotes the constant function. Moreover, any family of commutative rings is generic of bound $1_{\mathbb{Z}^+}$. A corollary of the product theorem is that any product $R = \Pi_{i\in I} R_i$ of a generic family of right FPF rings is right FPF. (In particular, the product any family of commutative FPF rings is FPF.) This implies that any product of self-basic right FPF rings, in particular, any product of self-basic right PF rings is right FPF.

Another corollary to the product theorem states that if $\{R_i\}_{i\varepsilon I}$ is any family of commutative rings each having the property $P(n,g)$: there exist integers $n \geq 0$ and $g > 0$ with the property that for all $i\varepsilon I$ every finitely generated R_i-module of free rank $\geq n+1$ has genus $\leq g$, then their product R also has property $P(n,g)$. (see Corollary 1.24) The FPF theorem is the case $P(0,1)$.

That a commutative ring is generic with bound the identity is given by:

<u>1.20A</u> <u>THEOREM</u>. [W. Vasconcelos].

<u>If</u> R <u>is a commutative ring, then</u> $\gamma(M) \leq \nu(M)$ <u>for any f.g. generator</u> M.

<u>Proof</u>. Let $M^n \twoheadrightarrow R$. Then there exist elements $x_1,\ldots,x_n \varepsilon M$, $f_1,\ldots,f_n \varepsilon M^*$ such that $\Sigma_{i=1} f_i(x_i) = 1$. If $t = \nu(M)$, and if $m_1,\ldots,m_t$ generate M, then $x_i = \Sigma_{j=1}^{t} m_j a_{ij}$ for some $a_{ij} \varepsilon R$, $i = 1,\ldots,n$. However,

$f'_j = \Sigma_{i=1}^{n} f_j a_{ij} \varepsilon M^*$, $j = 1,\ldots,t$ is such that $\Sigma_{j=1}^{t} f'_j(m_j) = 1$, so that $M^t \twoheadrightarrow R$ holds, that is, $\gamma(M) \leq t = \nu(M)$.

<u>1.20B</u> <u>COROLLARY</u>.

<u>If</u> M <u>is a f.g. faithful projective over a commutative ring</u> R, <u>then</u> (M <u>generates mod</u>-R <u>and</u>) $\gamma(M) = \gamma(M^*) \leq \nu(M)$.

<u>Proof</u>. M generates mod-R by a theorem of Azumaya [66].

For the proof of the product theorem we also need:

<u>1.21</u> <u>LEMMA</u>. [Faith 79c]

<u>The only f.g. right ideal</u> H <u>of a product</u> $\Pi_{i\varepsilon I}R_i$ <u>of rings which contains the direct sum</u> $\oplus_{i\varepsilon I}R_i$ <u>is the unit ideal</u>.

<u>Proof</u>. Let H be generated by elements $m^1,\ldots,m^t$, and for any $x\varepsilon R$, write $x_j = xe_j$, $\forall\, j\varepsilon I$. Since $e_j\varepsilon H$, $\forall\, j\varepsilon I$, there exist $a^{ij}\varepsilon R$, $i = 1,\ldots,t$, such that

$$e_j = \sum_{i=1}^{t} m^i a^{ij} = \sum_{i=1}^{t} m_j^i a_j^{ij}. \qquad (1)$$

Let $b^i\varepsilon R$ be such that $b_j^i = a_j^{ij}$, $\forall\, j\varepsilon I$. Then, clearly, the element

$$m = \sum_{i=1}^{t} m^i b^i \varepsilon M$$

is the unit element 1 of R since by (1)

$$m_j = \sum_{i=1}^{t} m_j^i b_j^i = e_j = 1_j$$

for any j. Thus, M is the unit ideal.

<u>1.22A PRODUCT THEOREM</u> [Faith 79c]

<u>A family</u> $\{R_i\}_{i\varepsilon I}$ <u>of rings is right</u> B-<u>generic iff the product</u> $R = \Pi_{i\varepsilon I}R_i$ <u>is</u> B-<u>generic</u>. <u>Thus, for every</u> $M\varepsilon\text{Gen } R$, <u>with</u> $\nu(M) = n < \infty$ we have:

$$\gamma(M) = \sup\{\gamma(M_i)\}_{i\varepsilon I} \le B(n)$$

where $M_i = Me_i$, <u>and</u> $e_i\varepsilon R_i$ <u>is the identity element</u>, $\forall\, i\varepsilon I$.

<u>Proof</u>. $M\varepsilon$ Gen $R \Rightarrow M_i\varepsilon$ Gen R_i for each $i\varepsilon I$; hence there are epics $M_i^\gamma \to R_i$, in mod-R_i, where

$\gamma = \sup \gamma_i \leq B(n)$; hence epics $h_i : M^\gamma \to R_i$ in mod-R. The image H of the product morphism $h : M^\gamma \to R$ satisfies $He_i = R_i$, $\forall i\varepsilon I$; hence H contains their direct sum, and Lemma 1.21 asserts that $H = R$. Thus

$$\gamma(M) \leq \gamma \leq B(n) = B(\nu(M)).$$

However, $\gamma(M) = \gamma$ since any epic $M^t \to R$ implies an epic $M_i^t \to R_i$, $\forall i\varepsilon I$.

Conversely, assume $R = \Pi_{i\varepsilon I} R_i$ B-generic, choose $i\varepsilon I$, and $M\varepsilon$f.g. Gen R_i. Let $n = \nu_i(M)$, and let $M^t \twoheadrightarrow R_i$, where $t = \gamma_{R_i}(M)$. Also let $N = \Pi_{j\neq 1} R_j$. Then $N_i \oplus M^t \twoheadrightarrow R = N \oplus R_i$, and hence $(N \oplus M)^{t'} \twoheadrightarrow R$, so $\gamma(N \oplus M) \leq t$. Note however that $(N \oplus M)^{t'} \twoheadrightarrow R$ would imply $M^{t'} \twoheadrightarrow R_i$, so actually $\gamma(N \oplus M) = t$. Moreover, $\nu_R(N\oplus M) = \nu_{R_i}(M) = n$, since:
$R^n = R \oplus R^{n-1} \twoheadrightarrow (N \oplus R_i) \oplus R_i^{n-1} = N \oplus R_i^n \twoheadrightarrow N$, (using $R \twoheadrightarrow N \oplus R_i = R$, and $R_i^n \twoheadrightarrow M$). Therefore, since R is B-generic, we have

$$t = \gamma_{R_i}(M) = \gamma_R(M) \leq B(n) = B(\nu_{R_i}(M)),$$

that is, $\{R_i\}_{i\varepsilon I}$ is B-generic.

It is clear from the proof that from the statement that $R = \Pi_{i\varepsilon I} R_i$ is a generic product of rings we may deduce either of the two equivalent properties:

(1) R is a generic ring (bounded, e.g., by B).

(2) $\mathfrak{F} = \{R_i\}_{i\varepsilon I}$ is a generic family (bounded, e.g., by B).

<u>1.22B COROLLARY.</u>

<u>If</u> M <u>is an f.g. module over a product of rings</u> $R = \Pi_{i\varepsilon I} R_i$, <u>if</u> $M_i = Me_i$ <u>generates</u> mod-R_i where $e_i : R \to R_i$

is the projection idempotent, and if $\sup\{\gamma(M)_i)\}_{i\varepsilon I} = \gamma < \infty$, *then* M *generates mod-R and* $\gamma(M) = \gamma$. *Thus*

$$\gamma(M) = \sup\{\gamma_{R_i}(M_i)\}_{i\varepsilon I}.$$

Proof. That M generates mod-R follows from the proof of the theorem which shows that if there exists $\gamma < \infty$ such that

$$\forall i\varepsilon I \ \exists \ M_i^\gamma \twoheadrightarrow R_i \text{ then } \exists \ M^\gamma \twoheadrightarrow R.$$

Moreover:

$$M^\gamma \twoheadrightarrow R \Rightarrow M_i^\gamma \twoheadrightarrow R_i.$$

hence (7.1) holds.

1.22C COROLLARY

Let $R = \Pi_{i\varepsilon I} R_i$. *Then*

$$g_r(R) = \sup\{g_r(R_i)\}_{i\varepsilon I} .$$

Proof. Follows from the corollary and the proof of the theorem.

1.22D COROLLARY

Any product of commutative FPF *rings is* FPF. *Similarly for products of right* FPF *self-basic rings*, (cf. 1.2B.)

Proof. Both are generic families.

1.22E COROLLARY

Any right generic product of right (F)PF *rings is right* FPF.

1.23 EXAMPLES

1.23A If F_n be the $n \times n$ matrix ring over a local ring

R, then the product $R = \prod_{n \in Z^+} F_n$ is not generic, since $\gamma(M) = \infty$ for the cyclic module $M = eR$, where $e = e^2$ is the idempotent the j-th component of which is the e_{11}-matrix in F_n.

<u>1.23B</u> An infinite product of right PF rings is never right PF since a semiperfect ring contains no infinite sets of orthogonal idempotents. Furthermore, a product of PF rings is not necessarily FPF, e.g., $R = \prod_{n \in Z^+} F_n$ is not. Nevertheless, any product of right PF rings of right genus $\leq g$ is right FPF of genus $\leq g$, according to Corollary 1.22E.

<u>1.23C</u> Let $R = \prod_{i \in I} M_n(F_i)$, where F_i is a self-basic right (F)PF ring. Then R is right FPF of genus n, according to Cor. 1.22E, since $G_r(M_n(F_i)) = n$, $i \in I$, by Example 1.23A.

<u>1.24 COROLLARY</u>

Let $R = \prod_{i \in I} R_i$ be a product of commutative rings such that there exists an integer $n > 0$ such that each R_i satisfies Serre's condition $P(n,g)$; that is, any finitely generated R_i-module of f rk $> n+1$ has a unimodular element. Then, R satisfies $P(n,g)$.

Proof. Let M be any finitely generated R-module of f rk $\geq n + 1$. If P_i is any maximal ideal of R_i, then $P = P_i \oplus R_i'$ where $R_i' = \prod_{j \neq i} R_j$, is maximal in R, and $(M_i)_P = M_P$ has rk $\geq n + 1$, so M_i has a unimodular element, that is, $\gamma(M_i) = 1$; hence $\gamma(M) + 1$ by Corollary 1.20B.

<u>1.25 THEOREM</u>

Let $\{R_i\}_{i \in I}$ be a family of rings such that R_i is a commutative ring of one of the following types:

(i) *a Bezout domain,*

(ii) *a local FPF ring (e.g., any* AMVR, *or any self-injective local ring),*

(iii) <u>an FPF ring of genus</u> 1,

(iv) <u>any product of rings</u> $\{R_1\}$ <u>where</u> R_1 <u>has type</u> (i)-(iv).

<u>Then</u>: $R = \Pi_{i \varepsilon I} R_i$, <u>is FPF of genus</u> 1;

<u>Proof</u>. The rings (i)-(iii) are all FPF of genus 1; hence by Corollary 1.22B so are the rings in (iv); hence so is $R = \Pi_{i \varepsilon I} R_i$.

A ring R (commutative) is said to be <u>quotient-injective</u> if its classical quotient ring $Q_{c\ell}(R)$ is a self-injective ring, equivalently, an injective R-module. Then R is said to be <u>fractionally self-injective</u> (FSI) if every factor ring of R is quotient-injective. Every FPF commutative ring R is quotient-injective, hence every CFPF commutative ring is FSI. Conversely, every FSI ring R is CFPF. (See Faith [77, 82] for these results, and the background). Now the FSI rings have been completely characterized by Vamos [77]: R is FSI iff R is a finite product of rings of the following three types: (1) AMVR; (2) Almost maximal h-local domain; (3) Almost maximal torch ring. Here, almost maximal means that every local ring of R is an AMVR; h-local means that every prime ideal P is contained in only finitely many maximal ideals; and a torch ring signifies that R is directly indecomposable (= has no non-trivial idempotents), has a minimal prime ideal P such that P is a uniserial R-module $\neq 0$, with $P^2 = 0$, and R/P of type (2). (Vamos [77,79] characterizes the FGC rings as Bezout FSI rings.)

This shows that no infinite product of rings can be CFPF, that is, the product theorem for FPF rings fails for CFPF rings. (Finite products of CFPF rings are CFPF however.)

See Faith [79c] for other results on the genus of modules and generic families.

NOTES FOR CHAPTER 1

We realize that Chapter 1 is difficult if not impossible to read, but thought that a road map placed at the beginning would only add to the confusion. So we summarize a bit here.

MORITA THEORY

1.1A Generators of mod-R

1.1B Definitions of (F)PF((FP^2F) and C(F)PF(C(F)P^2F) rings.

1.1C Similarity or Morita Equivalence of rings denoted $A \sim B$.

1.1D Morita's Generator Theorem.

1.2A Krull-Schmidt Theorem and Exchange Lemma.

SEMIPERFECT RINGS AND THE BASIC MODULE AND RING

1.2B The Basic Module and Basic Ring of a semiperfect ring.

1.2C Epic images of the Basic Module

1.2D-E (F)PF, and (F)P^2F, C(F)PF, C(F)P^2F are Morita invariant properties.

BOUNDED RINGS

1.3A-B FPF rings are bounded.

1.3C Products of bounded rings rings are bounded but not conversely.

1.3E Self-injective strongly bounded rings are FPF.

1.3F Completely self-injective duo rings are CFPF.

SERIAL AND QF RINGS

1.6 QF Rings are the Self-injective Artinian (Noetherian) Rings.

PF RINGS

1.7A PF Rings are the self-injective semiperfect rings with essential socles, equivalently injective cogenerator rings.

1.7B PF Rings have strongly bounded basic rings.

1.7C Kato's Theorem: Two sided cogenerators are two-sided PF.

1.8A (1) QF Rings are two-sided PF.
(2) Uniserial Rings are CFPF.

1.8B Osofsky's Theorem. Two-sided PF one-sided perfect rings are QF.

1.9 Tachikawa's Theorem. A left perfect right FPF ring is right PF.

1.10 Nakayama's Theorem. CPF Rings are uniserial.

1.11 Perfect CFPF Rings are uniserial.

FINITE GOLDIE OR UNIFORM DIMENSION

1.12A Goldie Dimension Theorem.

1.12B Semiperfect Rings of Finite Goldie Dimension.

NONSINGULAR MODULES AND RINGS AND MAXIMAL QUOTIENT RINGS

(Resume of Results)

ANNIHILATOR RIGHT IDEALS

(Resumé)

CLASSICAL QUOTIENT RINGS $Q_{c\ell}(R)$

1.12C Goldie's Theorem.

PRINCIPAL IDEALS RINGS

1.13 Basis Theorem for PIR's

MODULES FLAT OVER ENDOMORPHISM RING

1.18-9 The Camillo-Fuller Theorems: A ring R is (F)PF iff (finitely generated) faithful modules are flat as modules over their endomorphism rings.

THE GENUS OF A MODULE AND GENERIC FAMILIES OF RINGS

1.22A The product theorem: The product of a B-generic family is B-generic.

1.22D The product of commutative FPF rings is FPF.

1.25 Theorem: Products of i) Bezout domains, ii) local FPF rings, iii) FPF rings of genus 1 are FPF.

2 NONCOMMUTATIVE SEMIPERFECT AND SEMIPRIME (C) FPF RINGS

This chapter contains results on the right FPF rings which are semiperfect. The semiperfect rings have two properties which facilitate the inquiry: I) They have a basic ring and primitive idempotents; II) The Krull-Schmidt theorem holds for direct sums of principal indecomposables. The first theorem states that any semiperfect right FPF (or FP^2F) ring R is a direct sum of uniform right prindecs (= principal indecomposable right ideals), the basic ring R_0 is strongly right bounded (Compare Theorem 1.7B: any right PF ring has basic ring R_0 which is strongly bounded on both sides) and is a direct summand of every finitely generated (presented) faithful right module. Moreover, if R or $R/\text{rad } R$ is prime, then R is a full matrix ring A_n over a local ring A which is right FPF (FP^2F) and if R is prime right FPF, then A is a two-sided valuation domain (Corollaries 2.1C and 2.11E).

A semiperfect right self-injective ring R with strongly right bounded basic ring is right FPF (Theorem 2.2); moreover, the converse holds provided that for each right prindec eR, every element of $e(\text{rad}R)e$ is a zero divisor of eRe, e.g. if $\text{rad } R$ is nil (Corollary 2.2B). Although in general the basic ring of a semiperfect FPF ring is not a product of local rings, that of any semiperfect right CFPF ring R is. (See Theorem 2.5, and 2.7 for a converse.)

For semiperfect right and left FPF rings we show that if all one sided zero divisors are two sided zero divisors, then the classical and maximal quotient rings coincide (all four of them) are self-injective (Theorem 2.17)

We show that if the intersection of the powers of the Jacobson radical is zero, then right and left regular elements are regular (Theorem 2.18). Also, we show right FPF semiperfect rings contain the singular submodule of their injective hulls and that every finitely generated module contained in the injective hull and containing the ring is isomorphic to the ring (Theorem 2.11 and Corollary 2.13.).

Some results on Noetherian semiperfect FPF rings (e.g. they split into products of prime rings and QF rings) are found in Chapter 5.

A primitive right FPF ring is simple Artinian (Prop. 2.8), and, hence any semiprimitive right CFPF ring is a subdirect product of simple Artinian rings (Corollary 2.9).

The main objective concerning the right FPF prime rings is: are they semihereditary? The affirmative result for semiperfect rings already has been discussed. Also, two-sided FPF Noetherian prime rings are hereditary, as are Noetherian prime one-sided FPF rings, a result proved in Chapter 4, but the one-sided case is open for non-Noetherian rings.

<u>2.1A</u> <u>THEOREM</u> (Faith [76b])

<u>If R is a semiperfect right FP^2F (FPF) ring, then each right prindec</u> eR <u>is a uniform right ideal, hence</u>

$$(2.1) \qquad R = \bigoplus_{i=1}^{n} e_i R$$

<u>is a direct sum of uniform right prindecs</u> $e_i R$, $i = 1,\ldots n$. <u>Moreover, the basic module is isomorphic to a direct summand of any faithful finitely presented (generated) right</u> R-<u>module, and the basic ring</u> R_0 <u>is strongly right bounded.</u>

<u>Proof</u>. By the results <u>sup</u>. Theorem 1.2B, write (2.1) for right prindecs $e_i R$, $i = 1,\ldots,n$.

By Theorem 1.2B, the basic module B is isomorphic to a direct summand of any generator. Moreover, by the same theorem, if R is selfbasic, a cyclic module R/I generates mod-R iff $I = 0$. Thus, any f.g. nonzero right ideal I

must contain an ideal $\neq 0$, that is, R/I must be not faithful, since it is f.p. This proves that any selfbasic right FP^2F ring R is strongly right bounded, and hence the basic ring of a right FP^2F is strongly right bounded by Theorem 1.2C.

Next, by the structure of semiperfect rings, we can write $B = \bigoplus_{i=1}^{m} e_iR$, for right prindecs e_iR , $i = 1,\ldots,m$, with $e_iR \not\approx e_jR$, $i \neq j$. To prove these are uniform right ideals, that is, $I \cap K \neq 0$, for any two nonzero submodules of each e_iR , $i \leq m$, it is a n.a.s.c. that this holds for I and K finitely generated. Thus, assume that I and K are f.g. right ideals of, say, e_1R such that $I \cap K = 0$. Then $U = e_1R/I \oplus e_1R/K$ is f.p. Moreover, $(1-e_1)R \approx R/e_1R$ is also f.p., and hence so is $M = U \oplus (1 - e_1)R$. We shall prove that M is faithful, i.e., that M generates mod-R.

Hence assume that Q is an ideal that annihilates M. Then, Q annihilates every e_iR , $i \geq 2$, and also Q annihilates U , hence $e_1RQ \subseteq I \cap K = 0$, so Q annihilates B, which is faithful, so $Q = 0$. Since M is faithful, hence a generator, then by Theorem 1.2B, e_1R is isomorphic to a direct summand of M . Since e_1R is not isomorphic to any e_iR, $i \geq 2$, then the Exchange Lemma 1.2A implies that either $e_1R \approx e_1R/I$, or $e_1R \approx e_1R/K$. In the first instance, I splits in e_1R by projectivity of e_1R, and then indecomposability of e_1R implies that $I = e_1R$, hence $K = I \cap K = 0$. Similarly for $e_1R \approx e_1R/K$. This proves that e_iR is uniform, $i = 1,\ldots,m$. Since every e_iR, $i > m$, is isomorphic to one of e_jR, $j \leq m$, then every right prindec is uniform.

Note: Any right FPF ring is right bounded--see 1.3B. Strong right boundedness is certainly false for general semiperfect right FPF rings, e.g. for $R = F_n$, the ring of all $n \times n$ matrices over a field F, $n > 1$. Since F_n is simple, none of the prindecs e_iR can contain an ideal $\neq 0$.

Recall the definition of Goldie dimension (1.12A), denoted $\dim M$ (and $\dim R$) below.

2.1B COROLLARY

Let R be a selfbasic semiperfect right FPF (FP^2F) ring. Then, any finitely generated (presented) faithful right R-module M has Goldie dimension dim M $\geq$ dim R. Moreover, then M $\approx$ R iff dim M = dim R. In particular, any finitely generated (presented) faithful right ideal $\approx$ R is a principal right ideal $\approx$ R.

Proof. Immediate from Corollary 1.12A.

2.1C COROLLARY (Faith [77])

Let R be a semiperfect right FPF (FP^2F) ring with radical J. If R is semiprime or R/J is prime, then $R = \prod_{i=1}^{n} A_n$, the full n × n matrix rings over right FPF (FP^2F) local rings A_i. If R is prime, then A is a right and left valuation domain (right Ore domain)[1].

Proof. First suppose that R is prime. Then, the basic ring A of R is prime. However, since A is strongly right bounded by Theorem 2.1 this can happen only if $A = e_1Re_1$, and $R \approx (e_1R)^n$, for a right prindec e_1R, and integer n > 0. Clearly A is local. Since right FPF (FP^2F) is Morita invariant by Theorem 1.2C, then A is right FPF (FP^2F). Since A is prime, and strongly right bounded, then A is an integral domain, and the last corollary implies that every finitely generated (presented) right ideal is principal, which in a local ring means that A is a right valuation domain (right Ore domain in case of right FP^2F, since A is uniform by Theorem 2.1).

When R/J is prime, then R is an n × n matrix ring A_n over a local ring A, by the Wedderburn-Artin theorem (which states that R/J is a full n × n matrix

---

[1]Local FPF rings are characterized in Faith [79b]. It is easy to see that the converse holds. A right FPF domain is two-sided Ore (see Theorem 3.16), and then right VR implies left Goldie implies left VR (see Corollary 4.22B).

ring over a field) and a theorem of Bass [60] (which states that orthogonal idempotents of R/J, and hence $n \times n$ matrix units, lift (See Lambek [76], p.104; ART, p.162.)).

2.1D COROLLARY

If R is a semiperfect, selfbasic, right CFPF ring, then R is a right duo ring. Thus, any semiperfect right CFPF ring is similar to a right duo ring.

Proof. The second statement follows from Theorem 1.2B, and the first follows from 2.1A. (Since R/A is also semiperfect and selfbasic, any right ideal I must equal the annihilator A of R/I, since A is the largest ideal of R contained in I, and I/A therefore contains no ideal of R/A, that is, $I = A$ by 2.1A.)

2.2A PROPOSITION

Any semiperfect right selfinjective ring with strongly right bounded basic ring is right FPF.

Proof. This is 1.4.

2.2B PARTIAL CONVERSE (Faith [76b, I])

Any semiperfect right FPF ring with nil radical is right selfinjective.

Proof. Assume that R is selfbasic, $R = e_1R\oplus\ldots\oplus e_nR$, for mutually nonisomorphic right prindecs e_iR, $i = 1,\ldots,n$. It suffices to prove that $uR + e_iR = e_iR$, for any u in the injective hull of e_iR, $i = 1,\ldots,n$. Set $U = uR + e_1R$. Since $M = U + (1 - e_1)R \subset \hat{R}$ is faithful, and finitely generated, then by 2.1A, $M \approx R\oplus X$, for some $X \in$ mod-R. Since every $e_iRe_i \approx \text{End}\, e_iR_R$ is a local ring, $i = 1,\ldots,n$, then by the Krull-Schmidt theorem 1.2, e_1R is isomorphic to a direct summand of M. But since R is selfbasic, $e_1R \not\approx e_iR$, $i > 1$, hence e_1R is isomorphic to a direct summand of U. But U is uniform inasmuch as e_1R and its injective hull are uniform, and hence $U \approx e_1R$, so that $B = \text{End}\, U_R$ is a local ring $\approx e_1Re_1$. Therefore B has nil radical $Q \approx e_1Je_1$, and since the endomorphism

$f : U \to U$ induced by the isomorphism $U \to e_1R$ has zero kernel (via uniformity of U), then f cannot be nilpotent (or even a zero divisor), that is, $f \in Q$. This implies that f is a unit of B, hence $f(U) = e_1R = U$, as required. This proves that e_1R, similarly, e_iR, $i > 1$, hence R, is injective.

2.2C COROLLARY.

Let R be a semiperfect ring with radical J, and write $R = \bigoplus_{i=1}^{n} e_iR$ as in (2.1). If R is right FPF, and if (*) every element of e_iJe_i has nonzero right annihilator in e_iRe_i, $i = 1,\ldots,n$, then R is right selfinjective.

Proof. Same proof as 2.2B: if $fg = 0$ in B with $g \neq 0$, then $\ker f$ contains $gU \neq 0$.

2.2D COROLLARY.

A local right FPF ring R is right selfinjective iff $\operatorname{rad} R$ consists of zero divisors.

Proof. An element x in a right selfinjective ring R satisfies $x^{\perp} = 0$ iff there exists $y \in R$ such that $xy = 1$, which in a Dedekind finite ring, e.g. a semiperfect or local ring, implies $yx = 1$ (ART, p.85) so $\operatorname{rad} R$ consists of zero divisors.

This generalizes the Levy-Klatt [69] theorem for AMVR's.

2.2E THEOREM. (Tachikawa [79])

A left perfect right FPF ring is right PF.

Proof. By Bass's theorems on left perfect rings (Bass [60]; also ART, Chap. 22), R has nil radical and essential right socle, so R is right selfinjective by Theorem 2.2B, and hence right PF by Theorem 1.7A.

The fact that a local FPF ring need not be a VR (as noted in Example 9E, p.183, Faith 79b) lends significance to the next corollary: any local CFPF ring is a VR.

2.3 COROLLARY. (to THEOREM 2.1)

If R is a right CFPF local ring, then the right

ideals of R are linearly ordered, that is, R is a right VR.

Proof. Let A_1 and A_2 be two proper right ideals of R. By 2.1D, then A_1 and A_2 are ideals of R, and hence so is $A = A_1 \cap A_2$. Now the product $M = R/A_1 \times R/A_2$ is a faithful R/A-module, and hence, by the theorem, R/A is isomorphic to a summand of M. Since R is local, R/A_i is indecomposable, i = 1,2, and so the Krull-Schmidt Theorem 1.2A implies that $R/A \approx R/A_1$ or $R/A \approx R/A_2$. Suppose $R/A \approx R/A_1$. Then the annihilator of each module is A (resp. A_1), hence $A = A_1 \subset A_2$.

2.4 COROLLARY

A semiperfect, selfbasic, right CFPF ring is right σ-cyclic.

Proof. If M is finitely generated, then by the theorem, $M \approx R/A \oplus X$, where $A = \mathrm{ann}_R M$, and $X \in$ mod-R/A. We may define the projective cover dimension of M, p.c. dim(M) = n, where n is the number of indecomposable summands in any direct sum decomposition of p.c.(M) into a direct sum of indecomposable modules. By the Krull-Schmidt theorem, n is unambiguous and p.c.(M) has at most n nonzero summands in any direct sum decomposition. Thus, the isomorphism $M = R/A \oplus X$ implies that m = p.c. dim X < n. If $B = \mathrm{ann}_R X$, then $B \supset A$, and the p.c. dim of X defined over R/B is = m. Hence, by induction, X is a direct sum of at most m cyclic modules, which proves that M is σ-cyclic.

2.5 STRUCTURE THEOREM (Faith [77])

Any semiperfect right CFPF ring R is similar to a finite product of right duo right VR right σ-cyclic right CFPF rings. (The basic ring of R is right duo and right σ-cyclic and a finite product of right VR's.)

Proof. By Theorems 1.2C, 1.2D, 2.1A, we may suppose that R is selfbasic. Then by 2.1D, R is right duo. Thus, $e_1^{\perp} = (1-e_1)R$ is an ideal call it A_1. Then

$e_1RA_1 = 0$, and hence $A_1 = (e_1R)$, so $e_1R(1-e_1) = 0$. Thus, $e_1R = e_1Re_1$; similarly $e_iR = e_iRe_i \forall i$.

Since $e_iR = e_iRe_i$, and

$$Re_i = \sum_{j=1}^{n} e_jRe_i = \sum_{j=1}^{n} e_jRe_je_i = e_iRe_i,$$

we see that e_i is a central idempotent, $i = 1,\ldots,n$, therefore $R = e_1Re_1 \oplus \ldots \oplus e_nRe_n$ is a product of local rings, $e_iRe_i, i = 1,\ldots,n$. Moreover, each direct factor of a CFPF ring is itself CFPF. (see 2.11C.) Finally, each of the local rings is (right duo) right VR and right σ-cyclic by 2.3 and 2.4.

The next result generalizes the Auslander-Goldman-Michler Theorem which is the special case when R is prime and Noetherian, once we prove that R is Dedekind (Chapter 4, 4.10; See Michler [69].)

<u>2.6</u> <u>COROLLARY</u> (to 2.5)

<u>Any semiperfect right CFPF ring is a finite product of full matrix rings over right duo right</u> VR <u>right</u> σ-<u>cyclic right CFPF rings</u>.

<u>Proof</u>. Since the basic ring R_0 is right CFPF and semiperfect, then we may apply 2.5 to obtain that R_0 is a finite product of right VR's as stated in 2.5. By the Morita theory (see 1.2B) $R \approx \mathrm{End}_{R_0}(e_0R)$, where e_0 is the basic idempotent. The fact that projective modules over VR's (which are local rings) are free, and the fact that e_0R is finitely generated and projective over R_0, imply that R has the stated structure.

<u>2.7</u> <u>CONVERSE</u> (to 2.5)

<u>Any ring similar to a finite product of right</u> VR <u>right</u> σ-<u>cyclic right duo rings is right CFPF</u>.

<u>Proof</u>. As noted in the proof of 2.5, CFPF is a Morita invariant, that is, a category property, so we may assume that R is a finite product of right VR's. Moreover, CFPF is preserved under finite ring products. (The proof of this is postponed until 2.11C.) Thus, we may assume that R

is a right VR. If A is an ideal, and M is any finitely generated faithful right R/A-module, then

$$(1) \qquad M = R/A_1 \oplus \dots \oplus R/A_n$$

is isomorphic to a finite direct sum of cyclic modules R/A_i, $i = 1,\dots,n$, which, since R is right VR, can be chosen such that $A_1 \subset \dots \subset A_n$. Thus, since R is right duo, the A_i are ideals, and $A_1 = \text{ann}_R M = A$, so that (1) shows that M is a generator of mod-R/A. Therefore R is right CFPF.

In the main theorem of this section, we have assumed for a right CFPF ring R, that R is semiperfect, in particular, that $R/\text{rad } R$ is semisimple Artinian. Thus $R/\text{rad } R$ is a (finite) product of simple Artinian rings. Below, 2.9 shows that every semiprimitive right CFPF ring is a sub-direct product of simple Artinian rings.

2.8 PROPOSITION (Faith [76c])

A right primitive right FPF ring R *is semisimple Artinian*.

Proof. If M is a simple faithful right R-module, then

$$M^n \approx R \oplus X$$

for some integer $n > 0$ and module X, implies that R is semisimple along with M^n.

A ring R is *semiprimitive* iff R is a subdirect product of primitive rings or equivalently, if the intersection of the primitive ideals of R is zero. (R is semiprimitive iff the Jacobson radical of R is zero. Jacobson [64], or ART, Chapter 26.)

2.9 COROLLARY

A semiprimitive right CFPF ring R *is a subdirect product of simple Artinian rings*.

Proof. R is a subdirect product of semisimple

rings by 2.8, and any semisimple ring is a finite product of simple (Artinian) rings.

Note that R need not be semisimple, e.g. Z is not. However, the example of $\mathbb{Z}$ is illustrative; any semiprime, i.e. any semiprimitive right FPF ring with the a.c.c. on ideals is a finite product of prime rings (3.4). Thus, any directly indecomposable semiprime right FPF ring with a.c.c. is prime.

A right FPF domain is right and left uniform by 3.16 and 3.16B or 2.1A shows that a right FPF local ring is right uniform. By 2.6, any semiperfect ring R is right CFPF iff similar to a finite product of right σ-cyclic right duo right VR's. Any VR is right uniform, and the next result shows that the weaker assumption suffices for right FPF.

2.10 PROPOSITION

Any ring similar to a finite product of right σ-cyclic right duo right uniform rings R *is right FPF.*

Proof. We may assume that R is right σ-cyclic, right duo and right uniform. If M is a finitely generated R-module, then

$$M = R/A_1 \oplus \ldots \oplus R/A/_n \tag{1}$$

is a finite direct sum of cyclic right modules R/A_i, $i = 1,\ldots,n$. Since A_i is an ideal, $i = 1,\ldots,n$, we obtain that M faithful implies

$$\operatorname{ann}_R M = \bigcap_{i=1}^{n} A_i = 0.$$

whence for some i, $A_i = 0$, by uniformity, say, $A_1 = 0$, so R is a summand of M, making M a generator. This proves R is right FPF.

2.11 THEOREM. (Page [83/84])*

If R *is a semiperfect right FPF ring and* M *is a finitely generated submodule of the injective hull of* R,

---

*For commutative R, this holds more generally for *any* FPF R since $Q_c(R)$ is injective. For non-commutative self-basic rings, this is contained in Faith [76a]. See Corollary 2.1B.

<u>which contains</u> R, <u>then</u> $M \approx R$.

<u>Proof</u>. Let M be as stated in the hypothesis. Now M is faithful so $M = P \oplus X$ where $0 \neq P$ is projective. We know $P \approx \oplus \sum_{i=1}^{m} p_i R$ where $p_i R \approx e_j R$ for some e_j. Choose P so that m is maximal. We claim X contains no projective submodules. Suppose not, then there exists an $x \in X$ so that $xR \approx e_{ji}R$ for some e_{ji} where $e_{j1}e_0 = e_{j1}$. Form $X \oplus (e_0 - e_{j1})R = N$. Now N is faithful and for any map of $(e_0 - e_{j1})R$ to $e_{j1}R$, the image is in $e_{j1}J$. It follows that X generates $e_{j1}R$ and hence $X \approx e_{j1}R \oplus Y$ for some Y. This contradicts the maximality of m and establishes the claim. Next write $1 = p + x$ with $p \in P$ and $x \in X$. We claim $x \in Z_r(M_R)$. To see this note that $e_{ij} = pe_{ij} + xe_{ij}$ for each $i = 1,\ldots,k$, and $j = 1,\ldots \ell_i$. Now $xe_{ij}R \approx e_{ij}R$ and the kernel of the map $R \to xe_{ij}R$ given by left multiplication by xe_{ij} is $(1-e_{ij})R \oplus W$ where $W \subset e_{ij}J$. Therefore, since $e_{ij}R$ is uniform, this kernel is essential and $xe_{ij} \in Z_r(M)$ for each $i = 1,\ldots,k$, and any j. But $x = \sum_{k,j}^{n} xe_{ij}$ so the claim is justified. We now have that $p^\perp = 0$ since $x^\perp$ is essential. This means $pR \approx R$ and that the uniform dimension of P is the same as that of R. Of course this implies that $X = 0$ and hence that $M = P$ is projective. The next task is to show P is isomorphic to R. To this end we will show $P = N_1 \oplus N_2$ where $N_1 \approx \oplus \sum_{j=1}^{\ell_1} e_{1j}R$ and N_2 is a sum of projective indecomposables none of which is isomorphic to $e_{i1}R$. Since P is a generator we know $P \approx e_{11}R \oplus Y$ for some Y. Choose $N_1 \approx \oplus \sum_{j=1}^{m_1} p_j R$, $p_j R \cong e_{11}R$, so that $N_1 \oplus N_2 = P$ and N_2 does not contain a summand isomorphic to $e_{11}R$. We want to show $m_1 \geq \ell_1$. Next notice that $A = (\Sigma_{j,h>1} e_{hj}R)^\perp \neq 0$ and is a two sided ideal contained in $\Sigma_j e_{1j}R$. To see this A cannot be zero for $\Sigma_{h\neq 1,j} e_{hj}R$ cannot generate $e_{i1}R$, by the Krull

Schmidt theorem. Also, $A \cap e_{1j_1}R = 0$ for some $e_{1j_1}R$ but $e_{1j_2}R \cap A \neq 0$ violates the fact that $e_{ij_1}R \approx e_{ij_2}R$, so A is essential in $\sum_j e_{1j}R$. Now the uniform dimension of A is ℓ_1. Let $1 = n_1 + n_2$ where $n_1 \in N_1$ and $n_2 \in N_2$. $N_2A = 0$ for N_2 is a sum of projectives isomorphic to summands of $(e_0-e_{11})R$. This gives $A \subset N_1$ so the uniform dimension $N_1 \geq \ell_1$. It follows that $m_1 \geq \ell_1$. Notice next that N_1 cannot generate $e_{2j}R$, so N_2 must generate $e_{2j}R$. But then, as we just have seen, $N_2 = N_3 \oplus N_4$ where $N_3 \approx \sum_{j=1}^{m_2} p_{2j}R$ and $m_2 \geq \ell_2$, and $p_{2j}R \approx e_{21}R$ for all j. The obvious induction now gives $P \approx R \oplus X$, but $X = 0$ by the uniform dimension argument.

2.12 COROLLARY*

Let R be a semiperfect right FPF ring with injective hull E_R and $Q = Q^r_m(R)$. Then for each $q \in E$, we have $qQ + Q \approx Q$.

Proof. Trivially $qR+R$ is dense in $qQ + Q$. But the isomorphism $qR+R \approx R$ lifts to a Q-isomorphism of E. Also, under this extended isomorphism, q is sent to an element of R hence of Q and one is also in the image so that the image of $qQ + Q$ is Q.

2.13 COROLLARY

Let R be as in Corollary 2.12. Then

$Z_r(E) = Z_r(R)$.

Proof. Let $x \in Z_r(E)$. We have that $xR + R = dr$ for some $d \in Q$ with $d^\perp = 0$. Now $d = r_1 + xr_2$ and there is an r_3 such that $dr_3 = x$. Since $d^\perp = 0$, $r_3^\perp = x^\perp$ is essential and $r_3 \in Z_r(R)$ which is contained in J since $Z_r(R)$ contains no idempotents. So $x = r_1r_3 + xr_2r_3$ or $x(1-r_2r_3) = r_1r_3$ and

$$x = r_1r_3(1-r_2r_3)^{-1} \in R.$$

*For commutative R, this holds for a general FPF ring. See loc. cit., p. 2.10n.

For a ring S and a subring R of S, Faith [82] calls R a <u>sandwich subring</u> of S if rad $S \subset R$. In case $Q = Q_m^r(R)$ we have that R is a sandwich subring of Q. Moreover, for semiperfect right FPF rings we have:

2.14 COROLLARY. (Page [83/84])*

If R is semiperfect, right FPF and $E = Q_m^r(R)$, then $J(R) \supset J(E) = Z_r(R) = Z_r(E)$.

Proof. For a right self-injective ring we have $Z_r(E) = J(E)$ by Utumi [65, Lemma 4.1].

The next lemma points out the importance of having right regular elements left regular.

2.15 THEOREM. (Page [83/84])

Let R be a semiperfect right FPF ring. If all right regular elements are left regular, then the regular elements are units in $Q_m^r(R)$.

Proof. Let Q be the right injective hull of R. Let $\Lambda = \mathrm{Hom}_R(Q,Q)$. Then Λ is a Dedekind finite ring, i.e. $x y = 1 \to yx = 1$ in Λ, since it has no infinite sets of orthogonal idempotents, Jacobson [50]. Let $r \in R$ be such that $r^{\perp} = 0$. Then the map $x \to rx$ induces an isomorphism of Q, i.e. a unit of Λ which we will denote by r also. So for some $\lambda \in \Lambda$, we have $\lambda o r = r o \lambda = 1$, i.e. $\lambda(r) = \lambda(1)\, r = 1$. Now take $\theta \in \Lambda$ such that $\theta(1) = 0$. To show that $\lambda(1)$ is in $Q_m^r(R)$, we must show that $\theta(\lambda(1))r = 0$, see Lambek [66, prop. 1, p.94]. We have that $\theta(\lambda(1))r = 0$. Let $\theta(\lambda(1)) = y$. Form $yR + R$. Let the embedding of $yR + R$ into R be given by F. Then $f(y)$ is in R and $f(y)\, r = 0$, so $f(y) = 0$, i.e. $f(\theta(\lambda(1))) = 0$ so that $\theta(\lambda(1)) = 0$. Hence $\lambda(1)$ is $Q_m^r(R)$ and r is a unit in $Q_m^r(R)$ which is also Dedekind finite for the same reason Λ was.

2.16 PROPOSITION.**

If R is a semiperfect FPF ring with right regular

---

*For commutative R, this holds more generally for any FPF ring by a theorem of Faith [82a].

**For commutative R, this holds because $Q(R) = Q_c(R)$ is injective (loc. cit.)

elements left regular, then $Q_m^r(R) = Q_{c\ell}^{\ell}(R) = Q$, the right injective hull of R.

Proof. We wish to show every element of Q is of the form $a^{-1}b$ for some a and b in R with a regular. Let $q \in Q$, then $qR + R \subset dR$ for some d in Q with $d^{\perp} = 0$. But $dR \supset R$ so there exists an element a in R with $da = 1$. Now a is right regular so $d = a^{-1} \in Q_m^r(R)$ hence $q \in Q_m^r(R)$. Also, $q = a^{-1}b$ for some b in R which completes the proof.

2.17 THEOREM. (Faith [76a], Page [83/84])

If R is a semiperfect FPF (both sides) ring, with right regular and left regular elements regular, then $Q_m^r(R) = Q_m^{\ell}(R) = Q_{c\ell}^r(R) = Q_{c\ell}^{\ell}(R) = Q$ and Q is a right and left self-injective.

Proof. This follows directly from the right and left hand versions of theorem 2.15.

Remark . If R is a ring with no infinite sets of orthogonal idempotents and with right injective $Q_m^r(R)$ then right regular elements must be left regular so the conditions on regularity are clearly necessary in order that the maximal quotient rings be injective.

Next we show that many semiperfect right FPF rings do have right or left regular elements regular.

2.18 LEMMA. (Page [83/84])

Let R be a semiperfect right FPF ring. Let d be a right regular element of R. Then $\bigcap_{n=1}^{\infty} Rd^n \supset {}^{\perp}d$.

Proof. Let $xd = 0$, and form $F = R \oplus R$. Let $M = (d,x)R$ and consider $F/M = N$. We claim N is faithful. To see this suppose $(0,1)r \in M$. Then $(0,1) = (d,x)r_0$ for some r_0. But then $dr_0 = 0$, so $r_0 = 0$ and $xr_0 = r = 0$. This also shows $(0,1)R \cap M = 0$. Now $(0,1)R \cong R$ so there is a map f of N into Q such that $f(0,1)R = R$. N is isomorphic to a finitely generated submodule of Q which contains R. By Theorem 2.11 we have an epimorphism γ of N onto R. Now let $\gamma(1,0) = r_1$ and $\gamma(0,1) = r_2$ Then $R = r_1R + r_2R$. We claim $r_1 \in Z_r(R) \subset J$. We have $(d^2,0)R \subset M$ for $(d^2,0) = (d,x)d$. This means $r_1d^2 = 0$. But since $(d^2)^{\perp} = 0$, $R \cong d^2R$. Also d^2R is right essential in R because the uniform dimension of d^2R is the same as that of R. This gives $r_1 \in Z_r(R)$ so, since r_1R is small in R, $r_2R = R$, hence r_2 is a unit. Now $r_1d + r_2x = 0$ so $-r_2^{-1}r_1d = x$ and $x \in Rd$. We may repeat the above to $M_n = (d^n,x)R$ and $N_n = F/M_n$ for any n and hence that $x \in R\,d^n$ for all n.

<u>2.19</u> <u>THEOREM.</u> (Page [83/84])

<u>Let</u> R <u>be a semiperfect right and left FPF ring. If for each idempotent</u> e, <u>and element</u> $d \in eJe$ <u>such that</u> $d^{\perp} \cap eRe = 0$, <u>we have</u> $\bigcap_{n=1}^{\infty} Rd^n = 0$, <u>then right regular elements are regular</u>.

<u>Proof</u>. If d is right regular and d is right regular modulo the radical, J, then d is a unit. So we can assume d is not right regular modulo J. It follows that there is an idempotent e, so that $de \in J$. Now for primitive idempotents f and g, with $fR \cong gR$, if $df \in J$ then $dg \in J$, too. This means we can take the idempotent e so that $de \in J$ and if f is a primitive idempotent with $f = (1-e)f = f(1-e)$, then $df \in J$, and if g is a primitive idempotent with $ge = eg = g$, $gR \cong fR$. This last statement implies that $ed(1-e)$ and $(1-e)de$ are in $Z_r(R) \cap Z_\ell(R)$. Now we have $d = ede + (1-e)d(1-e) + z$ with $z \in Z_r(R) \cap Z_\ell(R)$. It follows that $ede + (1-e)d(1-e)$ is a right regular element. It is easy to see that $(1-e)d(1-e)$ is right regular in $(1-e)R(1-e)$ modulo $(1-e)J(1-e)$ since $ed(1-e) \in J$. So $(1-e)d(1-e)$ is a unit in $(1-e)R(1-e)$. Now apply lemma 7.1 to the right regular element $ede + (1-e)d(1-e)$ and any y such that

$$y(ede + (1-e)d(1-e)) = 0.$$

This says $y \in \bigcap_{n=1}^{\infty} R(ede + (1-e)d(1-e))^n =$ $\bigcap_{n=1}^{\infty} R((ede)^n + ((1-e)d(1-e))^n) = \bigcap_{n=1}^{\infty} R((1-e)d(1-e)^n$ in particular that $ye = 0$. But for $y = y(1-e)$, $y(1-e)d(1-e)u = y(1-e)$ for some $u \in (1-e)R(1-e)$ and hence that $y = 0$. We now have that $ede + (1-e)d(1-e)$ is regular. But since $z \in Z_\ell(R)$, it follows that d is left regular hence regular.

2.20 COROLLARY.

If R is right FPF and has a.c.c. on left annihilators, then right regular implies left regular.

Proof. Let $d^\perp = 0$ such that ${}^\perp d$ is maximal. Then lemma 2.18 implies that if $yd = 0$, then $y = rd$ for some $r \in R$. But then $rd^2 = 0$ and since ${}^\perp d$ is maximal ${}^\perp d = {}^\perp(d^2)$ and hence $rd = y = 0$.

2.21 COROLLARY. (Page [83/84])*

If R is left Noetherian right FPF, then R is a left order in a Quasi-Frobenius ring.

Proof. By Proposition 2.16 and Corollary 2.20 $Q^\ell_{c\ell}(R) = Q^r_m(R)$. This implies $Q^\ell_{c\ell}(R)$ is right self-injective and left Noetherian and therefore Quasi-Frobenius.

2.22 Corollary.

If R is right and left FPF and semiperfect, with $\bigcap_{n=1}^{\infty} Rd_1^n = \bigcap_{n=1}^{\infty} d_2^n R = 0$ for all right regular d_1 and left regular d_2, of eJe in eRe for any idempotent then,

$$Q^r_m(R) = Q^\ell_m(R) = Q^\ell_{c\ell}(R) = Q^r_{c\ell}(R) = Q.$$

Proof. This just combines theorem 2.17 and 2.19.

We obtain a partial converse namely,

---

*For commutative R, this is a theroem of Endo [68], and for Noetherian semiperfect, it derives from Faith [76c]. See Chapter 5.

2.23 THEOREM. (Page [83/84])*

Let R be a semiperfect ring, with i) $Q^r_m(R) = Q^r_{c\ell}(R) = Q = Q^\ell_{c\ell}(R)$, ii) $Q^r_m(R)$ is right FPF, iii) e_0Re_0 is strongly bounded, iv) every finitely generated right ideal of R which contains a regular element is a generator. Then R is right FPF.

Proof. Let M be a finitely generated faithful right R-module. We wish to show $M\ _RQ$ is a faithful Q module. It is easy to see $M\,Re_0 = N$ is a faithful e_0Re_0-module. Now let $\{n_i\}_{i=1}^k$ generate N over e_0Re_0. Since e_0Re_0 is strongly bounded $\bigcap_{i=1}^{\infty} n_i^{\perp} = 0$. So e_0Re_0 embeds in $N^{(m)}$ for some m. We have by Utumi, [65 p.219, prop. 3.2] that e_0Qe_0 is the maximal right ring of quotients of e_0Re_0. We claim e_0Qe_0 is left flat over e_0Re_0. To see this, we know Q is left flat over R because $Q = Q^r_{c\ell}(R)$ so $IQ = I\ Q$ for all right ideals I of R. Now for a right ideal H of e_0Re_0, H is of the form $H = e_0Ie_0$ for a right ideal I of R and
$H\otimes_{e_0Qe_0} e_0Qe_0 \cong e_0Ie_0\otimes e_0Qe_0 \cong e_0Ie_0\otimes_{e_0Re_0} e_0Q\otimes_{e_0Qe_0} Qe_0 \cong$
$(e_0Ie_0\otimes_{e_0Re_0} e_0R)\otimes_R(Q\otimes_{e_0Qe_0} Qe_0) \cong (e_0Ie_0\otimes e_0R\otimes_R Q)\otimes Qe_0 \cong$
$(e_0I\otimes_R Q)\otimes_{e_0Qe_0} Qe_0 \cong (e_0IQ)\otimes_{e_0Qe_0} Qe_0 \cong e_0IQe_0$ so that e_0Qe_0 is left flat over e_0Re_0. We have an exact sequence $0 \to Re_0 \to N^{(m)}$. Tensoring this with e_0Qe_0 over e_0Re_0 gives $0 \to e_0Qe_0 \to N^{(m)}\otimes e_0Qe_0 \cong (M\otimes Re_0)^m \otimes e_0Qe_0$. So $(M \otimes Re_0) \otimes e_0\,Qe_0$ is a faithful e_0Qe_0 - module. But then $M\otimes_R Q$ is a faithful Q-module. Now $M\otimes_R Q$ must generate Q. So there are maps $f_i: M\otimes_R Q \to Q$ so that $\sum_{i,j} f_i(m_j \otimes q_{ij}) = 1$. We have that the image of M in $M \otimes Q$ generates $M \otimes Q$. Also we can take the $\{m_j\}$ to generate M. Letting $f_i(m_j \otimes 1) = b_{ij}^{-1}a_{ij}$ and $q_{ij} = c_{ij}d_{ij}^{-1}$ we can find regular b and d so that $bf_i(m_j \otimes 1) \in R$ for all i and j and $q_{ij}d \in R$ for all i and j. Then each bf_i restricted to the image of M in $M \otimes Q$ gives a map of M into R and $bd = \Sigma\,(bf_i\,(m_j \otimes q_{ij}d)$ so bd is in the trace of M in R. By condition iv) M is a generator.

---

*For commutative R, this is contained in Faith [82a]. (In this case R is a product of local rings.)

3 NONSINGULAR RINGS

In this chapter we study nonsingular FPF rings. For commutative rings we have: If R is FPF, then $Q_c = Q_{c\ell}(R)$ is FPF, Faith [78, Th. C]. For general noncommutative rings the question remains open. In case R is right nonsingular or semiprime we show that $Q_m = Q^r_{max}(R)$ is FPF (both sides) if R is FPF (one side). In fact, under these circumstances we give a complete description of $Q_m(R)$ as a regular ring of bounded index. Using this description we are able to show prime right FPF rings are Goldie on both sides and that every right Goldie right FPF ring is left Goldie, too. We also show that right FPF rings are right semihereditary iff left semihereditary iff they and all their matrix rings are Baer rings. Finally we include some results on fully idempotent rings and in particular V-rings which are FPF.

Below and hereafter, ${}^{\perp}X$ (resp. $X^{\perp}$) denotes the left annihilator left (resp. right) ideal corresponding to a nonempty subset X of R.

<u>3.1A</u> <u>PROPOSITION</u>.

<u>Let</u> R <u>be right FPF</u>. <u>Then:</u>

3.<u>1A.1</u>

<u>If</u> M <u>is a finitely generated and faithful right R-module, then the dual module</u>

$$M^* = \mathrm{Hom}_R(M,R) \neq 0$$

3.1A.2

If I and K are right ideals such that $I \cap K = 0$, then ${}^{\perp}IR + {}^{\perp}KR = R$, and hence ${}^{\perp}I + {}^{\perp}K$ generates R-mod. Thus, either ${}^{\perp}I \neq 0$, or ${}^{\perp}K \neq 0$. Moreover, if I and K are ideals such that $I \cap K = 0$, then $R = {}^{\perp}I + {}^{\perp}K$.

Next let R be right FP^2F. Then:

3.1A.3

Both A.1 and A.2 hold with the proviso that M in A.1 is finitely presented, and I and K in A.2 are finitely generated.

3.1A.4

If R is local, then R is right uniform.

Proof. 1. Since M is a generator, by 1.1A, we have, for some integer n, and module X, an isomorphism $M^n \approx R \oplus X$. Thus, $M \neq 0$ follows from the isomorphism:

$$(M^*)^n \approx (M^n)^* \approx R \oplus X^*$$

Proof. 2. More generally, any direct sum decomposition $M = A \oplus B$ yields an isomorphism $M^* = A^* \oplus B^*$ in such a way that the trace T of M in R is the sum of the trace ideals of A and B, that is, the sum of the images of the canonical maps $A \otimes A^* \to R$ and $B \otimes B^* \to R$. Since $M = R/I \oplus R/K$ is faithful, then right FPF implies that M generates mod-R. Now $(R/I)^* \approx {}^{\perp}I$ canonically, so the trace of R/I in R is the image of the canonical map $R/I \otimes {}^{\perp}I \to R$, which is ${}^{\perp}IR$. Similarly, the trace of R/K is ${}^{\perp}KR$, so that $R = {}^{\perp}IR + {}^{\perp}KR$ is the trace of M. It follows therefore that ${}^{\perp}I + {}^{\perp}K$ is a generator of R-mod. Moreover, if I and K are ideals, then so are their left annihilators, so then $R = {}^{\perp}I + {}^{\perp}K$. In either case, ${}^{\perp}I$ and ${}^{\perp}K$ can not both vanish.

Proof. 3. The proofs go through *mutatis mutandis* for f.p. module M over a right FP^2F ring R. Furthermore, if I and K are f.g., then R/I and R/K are f.p., hence $M = R/I \oplus R/K$ is f.p., so the conclusions of 3.1A.2 also hold for f.g. I and K over a right FP^2F ring

R.

Proof. 4. This special case of 3.1A also follows from A.2 for f.g. I and K.

The part of the next theorem which is used in the sequel is the first statement which is a triviality. Therefore, although of peripheral interest, the rest may be skipped without interrupting the continuity.

An ideal which is an annihilator right ideal is said to be a *right-annihilator ideal*. Any right-annihilator ideal I is the annihilator $K^{\perp}$ of a left-annihilator ideal, that is, K can be chosen to be an ideal, in fact, $K = {}^{\perp}I$. A *right-inessential* (prime) ideal P is an (prime) ideal which is not an essential right ideal.

3.2 THEOREM.

(1) *In any ring* R, *any maximal right-annihilator ideal is a prime ideal. A partial converse: any prime right-annihilator ideal* P *is a maximal right-annihilator ideal if* P *does not contain a nilpotent ideal* $\neq 0$.

(2) *Any right-inessential prime ideal* P *is a right-annihilator ideal such that* $({}^{\perp}P)^2 \neq 0$, *and* P *is maximal in the set of right-inessential prime ideals.*

(3) *Any prime ideal* P *such that* $({}^{\perp}P)^2 \neq 0$ *is a minimal prime ideal.*

Proof. 1. The first assertion is obvious. Next, let P' be a right-annihilator ideal containing P. Since P contains no nilpotent ideals, then ${}^{\perp}P' \not\subset P$, hence $0 = {}^{\perp}P'P' \subset P \Rightarrow P' \subset P$, so $P = P'$.

2. If K is a right complement of P in R, then $K \neq 0$ since P is inessential, and $P \cap K = 0$ implies that $P \not\supset K$, so that ${}^{\perp}P \not\subset P$. Now $Q = P$ satisfies $0 = QQ^{\perp} \subset P$, and primeness of P implies that $Q^{\perp} \subset P$, and $Q^{\perp} = P$ is therefore a right-annihilator. Moreover, $({}^{\perp}P)^2 \neq 0$, since ${}^{\perp}P \not\subset P$. Maximality of P follows from (3).

3. Suppose that M is a prime ideal contained in P, and $Q^2 \neq 0$, where $Q = {}^{\perp}P$. Then, $0 = QP \subset M$, and $Q^2 \neq 0 \Rightarrow Q \not\subset P$ so $Q \not\subset M$, and hence $P \subset M$, that is, $P = M$.

Completion of the proof of (2). Let P' be a right-inessential prime ideal containing P. Then, by (3), P' is a minimal prime, hence $P' = P$, so that P is maximal as stated.

We consider the connection between semi-prime rings and right non-singular rings which are right FPF. For commutative rings non-singular and semi-prime are equivalent notions. If R is commutative and $x \in Z(R)$ = the singular ideal of R, then $(Rx) \cap ({}^{\perp}x) \neq 0$, but is square zero. So if R is semi-prime, then $Z(R) = 0$. Conversely, if x is in the prime radical of R and A is an ideal maximal w.r.t. $A \cap Rx = 0$, then $A \cdot Rx = 0$ and $A + Rx$ is essential. So if $Z(R) = 0$, evidently $x^2 \neq 0$, unless $x = 0$. For non-commutative R the notions are independent. For the class of right FPF rings we have:

3.3. THEOREM. (Page [82])

*Let R be a right FPF ring. Then $Z_r(R) = \text{sing}^r R = 0$ iff R is semiprime.**

Proof. Let $x \in Z(R)$ and form xR. If $(xR)^{\perp} = 0$, then xR is faithful and so would generate R, forcing R to be equal to $Z(R)$. So $(xR)^{\perp} = A \neq 0$. Now letting $B = xR \cap A$ gives $B^2 = 0$. But since R is semi-prime, $B = 0$, so $A \cap xR = 0$. Notice that, in fact, A is a right ideal maximal w.r.t. $A \cap (RxR) = 0$, since $A = (RxR)^{\perp}$ and if H is a right ideal such that $(RxR) \cap H = 0$, then $H(RxR) = 0$, but ${}^{\perp}(RxR) = (RxR)^{\perp}$ by semi-primeness, so $H \subset A$. Next, notice that $R/A \oplus R/RxR$ is faithful, hence generates R. This implies that ${}^{\perp}A + {}^{\perp}(RxR) = R$. But ${}^{\perp}A \cap {}^{\perp}(RxR) = {}^{\perp}A \cap A = 0$. $A = eR$ for e a central idempotent. We have that $eR \oplus xR$ is faithful, for if $(xR)y = 0$, $y \in eR$ and, $ey \neq 0$, unless $y = 0$. This means $eR \oplus xR$ generates R. That is, $\text{trace}(eR) + \text{trace}(Rx) = R$. Since $\text{trace}(eR) = eR$ it follows, because e is central, that $(1-e)R$ is generated by xR, i.e. $(1-e)R \subset Z(R)$, so that $e = 1$, and $Z(R) = 0$.

To prove the converse assume A is a two sided

---

*For some rings, e.g. commutative, this holds in general (without the FPF assumption!)

ideal with $A^2 = 0$. Let B be a right ideal maximal w.r.t. $B \cap A^{\perp} = 0$. If $B = 0$, then $A^{\perp}$ is essential and so $A \subset Z(R) = 0$. In case $B \neq 0$ and H is a two sided ideal with $H \subset B$, then $AH \subset B \cap A$, so $AH = 0$ and hence $H \subset A^{\perp} \cap B$. This means the only two sided ideal contained in B is zero. But this says R/B is faithful, so generates R. Now $A^{\perp}$ embeds as an essential submodule of R/B under the natural map, by the maximality of B. If f is a homomorphism from R/B to B, we have $f(\bar{1}) = b$. Letting A_1 be the image of A in R/B, $f(A_1) = bA^{\perp} = 0$, since $A^{\perp} \cap B = 0$. This says that each homomorphism from $R/B \to B$ has essential kernel, hence that B is contained in $Z(R)$.

3.3B COROLLARY

Let R be a right FPF ring with $Z_r(R) = 0$. Let A be a two sided ideal of R. Then A is essential in a two sided direct summand of R.

Proof. Let K be a right complement of A and B a right complement of A containing A. Then K and B are ideals such that $K = A^{\perp} = {}^{\perp}A$ and $B = K^{\perp} = {}^{\perp}K$. Then $R = B \times K$ by 3.1.

The next result reduces the structure of semiprime FPF rings with the a.c.c. on ideals to that of prime FPF rings.

3.4 THEOREM.

1. Any semiprime right FPF ring R with no infinite set of orthogonal central idempotents is a ring product of a finite number of prime right FPF rings.

2. A finite product $R = \Pi_{i=1}^{n} R_i$ of rings is right (C)FPF iff R_i is right (C)FPF), $i=1,\ldots,n$.

Proof. 1. Let P be any right annihilator ideal, and let $Q = {}^{\perp}P$. Then, semiprimeness of R implies that $Q = P^{\perp}$, and that ${}^{\perp}Q = P$ and that $P \oplus Q = R$. If R is prime, there is nothing to prove, otherwise we may suppose that $P \neq 0$. Since R is semiprime, $Q \cap P = 0$, and 3.1A.2 implies that $R = {}^{\perp}Q + {}^{\perp}P = P \oplus Q$. This is a ring direct sum decomposition since P and Q are ideals. Clearly P and Q inherit the hypotheses: if A is any finitely generated faithful right P-module, then $M = A \oplus Q$ is a finitely generated faithful right R-module, hence generates mod-R. However, as in the proof of 1A.2, there is an isomorphism $M^* = A^* \oplus Q^*$ in such a way so that the trace of M in R is the sum of the traces of A and Q. But, $Q = R/P$, so the trace of Q is ${}^{\perp}P = Q$, so it follows that $R = H + Q$, where H is the trace of A in R. Then $R = P + Q \Rightarrow H \supseteq P$, and hence the trace of A in P equals P, so A generates mod-P, that is P is right FPF. Also P has no infinite sets of central orthogonal idempotents, so we continue: either P is prime, or else P has a ring decomposition. By the nonexistence of infinite sets of central orthogonal idempotents, then R must have a prime factor, say $R = P_1 \times P_2$, where P_1 is a prime ring. Similarly, P_2 has a prime factor, and, by the nonexistence of infinite sets of orthogonal idempotents, we conclude that R has the stated decomposition into a finite product of prime rings.

2. The proof of 1. showed that if a ring product $R = P \times Q$ is right FPF, then so is P and Q, a result which trivially extends to any finite ring product $R = \Pi_{i=1}^{n} R_i$. Conversely, if R_i is right FPF, $i = 1,\cdots,n$, and if M is a finitely generated faithful right R-module, $M = Me_1 + \cdots + Me_n$ is a direct sum, where e_i is the identity element of R_i (= central idempotent), making $M_i = Me_i$ into a faithful finitely generated right R_i-module, and hence a right generator of mod-R_i, $i = 1,\cdots,n$. It easily follows that M generates mod-R; for if m_i is an exponent such that $M_i^{m_i} \to R_i \to 0$ is exact, $i = 1,\cdots,n$, then $M^m \to R \to 0$ is exact for $m = m_1 + \cdots + m_n$.

To prove that R is right CFPF iff R_i is, for $i = i,\cdots,n$, we first note that any ideal A of R has the form $A = A_1 + \cdots + A_n$, where $A \cap R_i = A_i$, $i = 1,\cdots,n$. Thus, $R/A = \Pi_{i=1}^{n} R/A_i$, and so the result for right CFPF follows from that for right FPF, if we note that every ideal of R_i is automatically an ideal of R. (Thus, for an ideal A of R_i, we have $R/A = R_1 \times R_2 \times \cdots \times R_i/A \times \cdots \times R_n$, and hence, if R/A is right FPF, then so is R_i/A, so that R right CFPF implies that of $R_i, i = 1, \cdots, n$.)

We are now in a position to prove:

<u>3.5</u> <u>THEOREM</u> (Page [82])*

<u>Let</u> R <u>be a right FPF ring with</u> $Z(R) = 0$. <u>Then</u> $Q(R)$ <u>is FPF</u>.

<u>Proof</u>. Letting $Q(R) = Q$ we take M a finitely generated Q-module which is faithful as a Q-module. Since R is essential in Q, the singular submodule of M, as a Q module, is the same as the singular submodule of M as an R-module. Suppose $M/Z(M)$ is not faithful, and let $A = (M/Z(M))^{\perp}$. A is a two sided ideal of R and by 3.3B A is essential in a direct summand, B, and R. If $m \in M/Z(M)$ such that $mB \neq 0$, then $mB \subseteq Z(M/Z(M)) = 0$. So A is a direct summand of R. Letting $D = {}^{\perp}A$ we see, much as in the proof of theorem 3.3A, that D is the right ideal maximal w.r.t. $A \cap D = 0$. But again D is essential in a direct summand of R, W say. Clearly $A \cap W = 0$ so $D = W$. It follows that $R = eR \oplus (1-e)R$ with e central, $e^2 = e$, $eR = A$. But then $M = Me \oplus M(1-e)$ with $Z(M) \subseteq Me$. Now eR is a FPF ring and is non-singular so we can assume that $R = eR$, $Q = eQ$, and $M = Me$ with $Z(M) = M$. We have exact sequences $Q^n \xrightarrow{\alpha} M \to 0$ and $R^n \to Q^n$, the later being the natural identifications. Let N be the image of R^n under α. Since N is a finitely generated R module and is singular, $A = N^{\perp} \neq 0$. But A is a two sided ideal and so is essential in a direct summand which is a two sided ideal, since ${}^{\perp}A$ is the right ideal maximal w.r.t. missing A, and A is essential in $({}^{\perp}A)^{\perp}$, a direct summand. Now our situation is that A is essential in eR, with e a central idempotent. If we take $N \oplus eR$, we have a faithful finitely generated module over R which generates R. As before N must generate $(1-e)R$, so $(1-e)R = 0$. But then

---

*For commutative R, the theorem holds for a general FPF ring, since $Q(R) = Q_c(R)$ is injective by Faith [82a].

A is essential in R and ${}^{\perp}A = 0$. We claim A contains a central idempotent. To see this we note that AQ is a right ideal of Q, and so $AQ \supset eQ \neq 0$ where $e^2 = e$. Let $e = \sum_{i=1}^{n} a_i q_i$, $q_i \in Q$ and $a_i \in A$. If $f: eR \to R$, then $f(e) = y \in R$ for some y. Since maps extend to Q, $y = ye$, and $\Sigma\, y a_i q_i = y$. It follows that $Ny = 0$ and hence $y \in A$. So trace$(eR) \subset A$. Trace(eR) is essential in an ideal B which is a direct summand. So $R = B \oplus C$ where $C = {}^{\perp}B = B^{\perp} = {}^{\perp}(\text{trace}(eR))$. Then $eR \oplus C$ is a faithful finitely generated module, hence generates R. It follows that trace$(eR) = B$. But B is generated by a central idempotent g, so trace$(eR) = gR$. But now $gR \subset A$, $NgQ = 0$ and gQ is a two sided ideal of Q. So $H = \{m \in M \mid mgQ = 0\} \supset N$ is an essential Q-submodule of M. We have a contradiction, since $H = M(1-g)$, unless $A = 0$. This implies R is singular which it is not so we can assume $M/Z(M)$ is faithful.

Now let M be any finitely generated faithful non-singular Q-module. Then for some n, $Q^n \xrightarrow{\alpha} M \to 0$ is exact. Let $R^n \to Q^n$ be the canonical embedding and let N be the image of R^n under α. We claim N is R faithful. If not let A be the two sided ideal which annihilates N. As above, A is essential in a two sided direct summand, so A is a direct summand, since $Z(N) = 0$. Then $A = eR$, where $e^2 = e$ is central. Then $NeQ = 0$, for e is central in Q, too. Because eQ is a two sided ideal in Q, the left annihilator of eQ is not zero, and is a Q-submodule of M. Since it contains N it must be Q essential, i.e., $M(1-e)$ is Q-essential in M. It follows that $e = 0$ and N is R faithful. Then N generates R, so $N^{\ell} \to R \to 0$ is exact for some integer ℓ. Since maps of N to R lift to maps of M to Q we have a Q-map, $M^{\ell} \to Q$, which must be an epimorphism since the identity is in the image.

<u>3.6</u> <u>THEOREM</u> (Page [82])

<u>If</u> R <u>is a ring with</u> $Z_r(R) = 0$ <u>and</u> R <u>is right FPF, then the left singular ideal of</u> R <u>is zero.</u>

Proof. If H is an essential left ideal of R and $0 \neq x \in H^{\perp}$, then $0 \neq xQ = eQ$, where $e^2 = e \in Q$. But then ${}^{\perp}(eQ) = Q(1-e) \supseteq H$ and H is not essential for $R(1-e) \cap R$ is not essential. To see this, clearly $ReR \cap R \neq 0$ and is essential in B, where B = trace eR, as in the previous proof. We also have $B \oplus C = R$ for some two sided ideal C. Now if $f \in \mathrm{Hom}(eR,R)$, $f(e) = y \in R$, and since maps extend to Q, $y = ye \in Re \cap R$. Because $B \neq 0$, $Re \cap R \neq 0$.

Over right nonsingular rings, with respect to a module M being a generator, the singular submodule has the fundamental property, used above, stated in 2. of the next proposition. Moreover, $M/\mathrm{sing}\, M$ is nonsingular, for any module M over a nonsingular ring R (e.g. ART, p.88, 19.46B).

3.7A. PROPOSITION

1. *If* R *is any ring, then a module* M *can be a generator only if* $\mathrm{sing}\, M \neq M$.

2. *If* R *is right nonsingular, and if* K *is a submodule of* $\mathrm{sing}\, M$, *then* M *is a generator of* mod-R *iff* M/K *is a generator*.

3. *If* R *is right nonsingular, then a module* $M \neq \mathrm{sing}\ M$ *iff* $M' = \mathrm{Hom}_Q(M,Q) \neq 0$.

Proof. 1. Let M generate mod-R. Then, there exists an integer $n > 0$ and a module X such that

$$(1) \qquad M^n \approx R \oplus X$$

Now $\mathrm{sing}\ M^n = (\mathrm{sing}\ M)^n$, so that $\mathrm{sing}\ M = M$ would imply $\mathrm{sing}\ M^n = M^n$, and then $\mathrm{sing}\ R = R$ a patent contradiction. This proves 1.

2. Since $M \to M/K$ is epic the if part is clear. Conversely, $K \subseteq \mathrm{sing}\ M \Rightarrow K^n \subseteq \mathrm{sing}\ M^n$ so if M generates mod-R, there is an isomorphism (1), call it h, and nonsingularity of R implies that

$$h(\mathrm{sing}\ M^n) = \mathrm{sing}\ (R \oplus X) \subseteq X,$$

hence

$$(M/K)^n = M^n/K^n \approx R \oplus (X/Y)$$

where $Y = h(K^n)$, so M/K generates mod-R by 1.1A.

3. Since Q is nonsingular, clearly $M' \neq 0$ implies that M has a nonsingular epimorph, consequently $M \neq S = \text{sing } M$. Conversely, M/S is n.s., and $(M/S)' \neq 0$ implies $M' \neq 0$, so it suffices to assume that M is n.s. If $x \in M$, $x \neq 0$, then x has inessential right annihilator I in R, so the double complement $\bar{I}$ of I is $\neq R$. Now a complement right ideal $\bar{I}$ has the form $e^{\perp} \cap R$ for an idempotent $e \in Q = \hat{R}$, where $(1-e)Q$ is the injective hull of I (of $\bar{I}$) in Q, and therefore $R/\bar{I} \approx eR \subset Q$ canonically. Now $xR \approx R/I$ together with the canonical map $R/I \to R/\bar{I} \approx eR \to Q$ yields a nonzero map $xR \to Q$ which by injectivity of Q extends to a nonzero map $f: M \to Q$.

3.7B THEOREM

A ring R in which any finitely generated right ideal $\neq 0$ generates mod-R, is right nonsingular.

Proof. Any finitely generated right ideal $J \neq 0$ generates mod-R, hence sing $J \neq J$ by 3.7A1. Thus, J is not contained in sing R, so sing $R = 0$. The application to prime right FPF rings is clear.

It seems reasonable at this stage to see what can be said about the regular FPF rings. In order to do so we introduce the following relevant concepts:

1. A ring R is of *bounded index* if there exists an integer $N > 0$ such that if $x^n = 0$, then $x^N = 0$.

2. Let M and N be R-modules. Let N-dim M denote the maximum of all integers n so that N^n embeds in M. Also, let

$$D(M) = \sup\{N\text{-dim } M \mid N \in R\text{-mod}\}.$$

The following result of Utumi [57] gives the connection between rings of bounded index and FPF rings.

We include the proof for completeness.

3.8A. THEOREM (Utumi [57])

Let R be a ring with zero singular right ideal. Then R is of bounded index if $D(R) < \infty$; and in case R is regular $D(R)$ equals the smallest bound on the index of nilpotence.

Proof. We can suppose R is regular for the maximal ring of quotients, $Q(R)$, is regular and R is an essential submodule of $Q(R)$. Suppose $x^n = 0$ but $x^{n-1} \neq 0$, for some $x \in R$. Let $K_1 = (x^{n-1})^\perp$ and consider $0 \to K_1 \to R \xrightarrow{x^{n-1}} x^{n-1}R \to 0$. The sequence splits by regularity of R, so $R \supset W_1 \approx x^{n-1}R$ and $W_1 \cap K_1 = 0$. Let $K_2 = \{x^{n-2}\}^\perp \cap xR$ and form $0 \to K_2 \to xR \to (x^{n-1})R \to 0$ which also splits. Therefore there exists $W_2 \subset xR$ with $W_2 \cap K_2 = 0$ and $W_2 \approx (x^{n-1})R$ so that $W_2 \approx W_1$. Also since $K_1 \cap W_1 = 0$ and $xR \subset K_1, W_2 \cap W_1 = 0$. By $n - 1$ applications of the above technique we obtain $W_1 \approx W_2 \approx \cdots \approx W_{n-1}$ with $(x^{n-1})R \subset K_i = \{x^{n-i}\}^\perp \cap x^i R$, and $W_i \cap K_i = 0$. It follows that $D(R) \geq n$ since $(\oplus \sum_{i=1}^{n-1} W_i) \oplus (x^{n-1})R \subset R$.

Next suppose $\{L_i\}_{i=1}^n$ is an independent set of right ideals in R with $L_i \approx L_j$ for all i and $j \leq n$. Since R is regular we can assume the L_i are all idempotent generated, by $e_1, e_2, \ldots, e_n$, say with $e_i e_j = 0$ for i, $j = 1, \ldots, n$, $i \neq j$. Let $\phi_{ij} : e_i R \approx e_j R$. Then ϕ_{ij} is left multiplication by $e_i r_{ij} e_j$ for some $r_{ij} \in R$. Let $x = \sum_{i>i} e_i r_{ij} e_i$. Then $x^n = 0$ but $x^{n-1} \neq 0$.

3.8B COROLLARY

If R is a domain which is not a right Ore domain, then the maximal right quotient ring $Q(R)$ is of unbounded index.

Another fundamental result is the following of Bumby [65], which generalizes the Cantor-Shröder-Bernstein theorem for sets. (See ART forward)

3.8C PROPOSITION (Bumby [65])

Let M_1 and M_2 be injective modules with exact sequences $0 \to M_1 \to M_2$ and $0 \to M_2 \to M_1$. Then $M_1 \approx M_2$.

We start with commutative rings, i.e., rings for which $D(R) = 1$, then using Morita equivalence build up to the more general case.

The next theorem is a special case of the main theorem of Faith [82a] which implies more generally that a commutative ring R is self-injective iff R is FPF and every regular element is a unit.

3.9A THEOREM

The following are equivalent for a commutative regular ring R.

i) R is self-injective

ii) R is FPF

iii) The trace of every finitely generated faithful module is finitely generated.

Proof. If R is injective and M is a finitely generated faithful module, then R embeds in a finite direct sum of copies of M as a direct summand. This gives i) $\Rightarrow$ ii).

That ii) implies iii) is trivial.

iii) $\Rightarrow$ i) Let $q \in Q$, the injective hull of R. Form $Rq + R = M$. Now trace(M) is finitely generated since M is finitely generated and faithful. Since R is regular and trace(M) is finitely generated, we have that trace(M) = eR, $e^2 = e$. Let $i \in I = \{r \in R | qr \in R\}$, an essential ideal. Then multiplication by i defines a map of M into R and this map sends 1 into i so $I \subset$ trace(M). Now take $f: M \to R$. Let $f(q) = x_0$ and $f(1) = y_0$. Then for every $z \in I$ we have $f(zq) = zqy_0$ so $z(x_0 - qy_0) = 0$, hence $x_0 = qy_0$ and $y_0 \in I$. I is generated by idempotents so we can take $y_0 = y_0^2$ so that $x_0 = x_0 y_0$, that is, trace(M) = I, too. Since I = eR and I is essential, $I = R$ and hence $q \in R$. This proves that (iii) $\Rightarrow$ (i).

3.9B COROLLARY

If R is a strongly regular ring (all idempotents are central), then R is FPF iff R is selfinjective.

Proof. If R is strongly regular, left ideals are ideals and are generated by idempotents. Also if M is finitely generated by $x_1,\dots,x_n$ say $M = \bigcap_{i=1}^{n} \{x_i\}^{\perp}$ for strongly regular rings. With these observations the previous proof goes through.

In order to obtain some insight into the connection with D(R) consider the following:

If D is a division ring and $R = \text{End}_D(V)$, then R is FPF iff V is finite dimensional over D, but R is always self injective and regular. The significant observation is that if V is infinite dimensional over D and $f \in R$ is a map with one dimensional range, then fR is finitely generated and faithful but can not generate R because, roughly, R contains infinitely many copies of fR i.e. fR - dim R = ∞. (Expressed otherwise, R is not semisimple.)

3.9C PROPOSITION

Let $\{R_i\}_{i\varepsilon I}$ be a collection of rings. Let $R = \Pi_{i\varepsilon I}R_i$ as rings. Then R is right FPF iff each R_i is right FPF and for each collection $\{M_i\}$ where M_i is a finitely generated faithful R_i-module $\forall i\varepsilon I$ such that $\Pi_{i\varepsilon I}M_i$ is a finitely generated R module, there exists an integer $N > 0$ such that R_i is a homomorphic image of a direct sum of N copies of M_i for each $i\varepsilon I$.

Proof. Routine coordinate-wise computation yields the proposition.

The previous proposition points out that if R is a product of matrix rings over division rings in order that R be right FPF the matrix rings had better not become too "large". It also suggests we look at the types given by Kaplansky [68] as refined by Goodearl and Boyle [76]:

1. *A regular right self-injective ring R is called type I if for every direct summand eR of R, $eR \supset H \neq 0$, a right ideal, such that for any right ideals $A \neq 0$ and $B \neq 0$ contained in H, $\text{Hom}(A,B) \neq 0$. If*

eR = H, e is called abelian.

2. A ring R is called Dedekind finite if xy=1 implies yx=1, otherwise we say R is Dedekind infinite.

3. A regular right self-injective ring R is called type II if R contains an idempotent e such that eR is faithful, eRe is Dedekind finite but R contains no abelian right ideals.

4. A regular right self-injective ring R is type III if $0 \neq e^2 = e$ implies eRe is not Dedekind infinite.

Type III rings are characterized by the fact that for any direct summand, H, then $H \approx H \oplus H$.

3.10 THEOREM [Kaplansky [68], Goodearl-Boyle [76, corollary 7.7 p.48]]

If R is a regular right self-injective ring, then $R = \Pi_{i=1}^{5} R_i$, where R_1 is type I and Dedekind finite, R_2 is type I and Dedekind infinite, and R_3 is type II and Dedekind finite, R_4 is type II and Dedekind infinite, and R_5 is type III.

Remark: All type III rings are Dedekind infinite. Also, we will adopt Kaplansky's [68, p.11] notation and say R is type I_f if R is type I and Dedekind finite, type I_∞ if type I and Dedekind infinite, type II_f and II_∞ are defined analogously.

3.11A. PROPOSITION

If R is a right self injective regular and right FPF ring, then R is biregular.

Proof. Let $x \in R$. We wish to show RxR is generated by a central idempotent. Let $H = (RxR)^{\perp}$. If H = 0, then xR generates so RxR = R. If $H \neq 0$, then H is the right ideal maximal with respect to $H \cap RxR = 0$. It follows that H is a direct summand of R because R is self-injective. Now $H \oplus xR$ is a finitely generated faithful module, hence a generator, so trace $(H \oplus xR) = H \oplus RxR = R$.

3.11B. PROPOSITION (Page [78])

If R is a right self injective regular right

FPF ring, then R is Dedekind finite.

Proof. If not, then by Goodearl, Boyle [76, Prop. 7.4 p.48] $R = R_1 \times R_2$ with $R_2 \neq 0$ and purely infinite, i.e. for every central idempotent $0 \neq e$, in R_2, eR_2e is not Dedekind finite. So assume $R \neq 0$ and purely infinite.

By Goodearl, Boyle [76,Thm. 6.2 p.41] there is in R a sequence of idempotents $e_1, e_2, \ldots$ such that for each i, $e_iR \approx R$, and $\sum_{i=1}^{\infty} e_iR$ is direct and essential in R. Let $M = R/\sum_{i=1}^{\infty} e_iR$. We claim M is faithful. If not, there exists $x \in R$ such that $RxR \subset M^{\perp}$. By proposition 3.11A, $RxR = eR$ for some central idempotent e. Since $Me = 0$ it follows that $eR \subset \sum_{i=1}^{\infty} e_iR$. But then $eR \subset \sum_{i=1}^{N} e_iR$ for some N large enough. This implies $eR \cap e_iR = 0$ for $j > N$, which implies $e_je = 0 \; \forall j > N$ since e is central. However, since $e_iR \approx e_jR$ for all i and j and e is central, then $e_ie = 0$ for all i, a contradiction.

Thus M is faithful; M is also singular, hence R is singular so must be zero.

3.11C. COROLLARY

If R is regular FPF type I, then R is of bounded index.

Proof. By Goodearl-Boyle [76, p.30] we see that if R is type I, R contains an indempotent such that eRe is strongly regular and eR is faithful. It follows that R is Morita equivalent to a strongly regular ring. Then using Tominaga [55, lemma 1, p.139] we see that R is of bounded index.

3.11D. PROPOSITION. (Page [78])

Let R be a regular right self injective right FPF ring of type II_f. Then $R = \{0\}$.

Proof. Let $0 \neq R$ be as above. We claim R can not be a simple ring. If R were a simple ring since it is type II it cannot be a semi-simple ring, hence must have an essential right ideal E. But then R/E is faithful by the simplicity of R hence a generator of R. This says $Z(R) = R$, a contradiction. Since R is not simple there must exist an idempotent $e_1 \varepsilon R$ such that $0 \neq Re_1R \neq R$. Now let $H_1 = (Re_1R)^{\perp}$. If $H_1 = 0$ then e_1R generates R which it doesn't, so $H_1 \neq 0$. Now H_1 is the right ideal maximal with respect to $H_1 \cap Re_1R = 0$, so $H_1 \oplus$ is a summand by injectivity of R. It follows that $H_1 \quad Re_1R = R$ as above. Now H_1 and Re_1R are type II_f right FPF rings so we can repeat the process to Re_1R to obtain an ideal $H_2 \subset Re_1R$. Continuing in this way we obtain $H_1 \oplus H_2 \oplus \ldots \subset R$ so that each H_i is a nonzero two sided direct summand of R. Since each H_i is type II_f we can choose an idempotent $f_i \varepsilon H_i$ such that $H_i = \oplus \sum_{j=1}^{i} f_{i_j} R$, $f_i R \approx f_{i_j} R$ for all $j \leq i$. Next take $gR = E(\oplus \Sigma H_i)$. We have that gR is a two sided ideal for the hull of any two sided ideal in a semiprime right self injective ring is complemented by its right annihilator which is a two sided ideal. We can assume then that g is a central idempotent. Form $\Pi_{i=1}^{\infty} f_i R$ and let M be the cyclic submodule generated by $((f_i)_{i\varepsilon I})R$. Let $N = M \oplus (1-g)R$. Then $Ny = 0$ iff $y(g-1) = 0$ and $f_i Ry = 0$ for all i, so $Rf_i Ry = 0$ for all $i \varepsilon I$. Then $(\sum_{i=1}^{\infty} H_i)y = 0$. But since $y = gy$ there exists an essential right ideal E such that $yE \subset \sum_{i=1}^{\infty} H_i$ and $(yE)^2 = 0$ implies $y = 0$ so N is faithful. Since R is right FPF, N generates R so $((f_i)_{i\varepsilon I})R$ must generate gR. It follows that for a fixed $n > 0$ there are maps making $\sum_{j=1}^{n} f_{i_j} R \to H_i \to 0$ exact for every i. But if $i > n$ we see by Bumby's result $H_i \oplus f_i R \approx H_i$ and R is not Dedekind finite.

Putting the above facts together gives:

3.11E THEOREM (Page [78])

A regular right self-injective ring is right FPF iff it is of bounded index.

3.11F. COROLLARY

A regular right self-injective ring is right FPF iff it is Morita equivalent to a strongly regular two-sided self-injective ring.

3.11G. LEMMA.

Let R be a right nonsingular right FPF ring and let $q \in Q$. Then $R + qR$ embeds in a finitely generated free module.

Proof. An idempotent e in Q is called abelian if for R-submodules I and J of eQ such that $I \cap J = 0$, $\mathrm{Hom}_R(I,J) = 0$. Now each idempotent of Q can be written as a finite sum of orthogonal abelian idempotents because Q is a self-injective regular ring of bounded index. The injective hull of qR is eQ for some idempotent e. Let $e = \sum_{i=1}^{n} e_i$, where the e_i's are abelian and orthogonal. Clearly, $R + qR$ embeds in $(1-e)R \oplus \sum_{i-1}^{n}(e_iR + eqR)$. Next look at $e_iR + eqR \subset e_iQ$. We will show that $Re_iR + ReqR$ embeds in a free module for each i. To this end, for convenience, we will assume e is abelian. Now we can reduce to the case where eR is faithful. To do this note that the right annihilator of $eR + eqR$, $(eR + eqR)^{\perp}$, is $(R(eR + eqR))^{\perp}$, a two sided ideal. Now 3.1A.2 implies that $R \approx R_1 \times R_2$ where $R(eReqR)$ is essential in R_1. We can, therefore, assume without loss of generality that $R = R_1$. This makes eR faithful and so $M = eR + eqR$ is a generator. This gives the existence of functions $f_1, \ldots, f_K, \in M^*$ so that $R = \sum_{i=1}^{K} \mathrm{Im}\, f_i$. Let $W = \cap_{i=1}^{K} \ker f_i$. Let F be the direct sum of K copies of R, and $Q(F)$ the canonical hull of F. Let f be the map of M to F given by f_i on the ith coordinate. We have $W = \ker f$. Since F is nonsingular, W is not essential in M. Choose U so that $W \oplus U$ is essential in M. Since $1 \in \sum_{i=1}^{K} \mathrm{Im}\, f_i$, there exists r_1, r_2 in R so that for $w \neq 0$ in W,

$(er_1 + eqr_2)f_i w \neq 0$ for some i. Also since the image of U is essential in im f, we see that $f(U)W \neq 0$, in Q(F). It follows, because all modules under consideration are nonsingular, that for some nonzero submodule $W_1 \subset W$, $\text{Hom}_R(W,U) \neq 0$, which contradicts the fact that e was abelian, unless W = 0. But W = 0 implies that the f_i's give rise to an embedding. Finally, treat (1-e)R in the same way.

3.12 THEOREM. (Page [81])

Let R be a right nonsingular right FPF ring. Then Q is flat as a right R module and $Q \otimes_R Q \approx Q$, that is, $R \hookrightarrow Q$ is a flat epi.*

Proof. Lemma 3.11G gives the essential ingredients to apply the proof of theorem 5.17 of Goodearl [76].

3.13A. PROPOSITION. (Page [81])

Let R be a regular right FPF ring. Let $e = e^2 \varepsilon Q$. Then eR is a projective R module.*

Proof. By theorem 2.8 of Sandomierski [68] it suffices to show $eR \otimes_R Q$ is a Q projective. Now we have $0 \to eR \to Q$ exact and Q is flat over R, so $0 \to eR \otimes Q \to Q \otimes Q$ is exact. The isomorphism $Q \otimes Q \approx Q$ gives $eR \otimes Q \approx Qe$, and hence is Q projective.

3.13B. COROLLARY.

For any idempotent $e \varepsilon Q$, $eR \cap R$ is a direct summand of R.*

Proof. The sequence $0 \to eR \cap R \to R \to (1-e)R \to 0$ splits.

We now have for regular FPF rings the following: If H is a right ideal of R, then H is essential in a summand e of Q. Hence H is essential in eR, hence essential in $eR \cap R$, a summand of R. This gives

3.13C. PROPOSITION.

If R is a right FPF regular ring, then R is right self-injective.

---

*For commutative R, the theorem holds more generally for any FPF ring. See Faith [79b, 82a].

Proof. If R is regular then certainly $Z(R) = 0$ and by corollary 3.13B each right ideal is essential in a direct summand of R. In regular rings each finitely generated right ideal is a direct summand (ART, Chap. 11), hence a right ideal isomorphic to a direct summand is a direct summand. These two properties constitute the definition of right continuous rings and the last corollary of Utumi [11, corollary 8.4] states that if R and any matrix ring over R are both continuous, then R is self-injective. Since both FPF and regularity are easily checked to be Morita invariant properties, it follows that R is right self-injective.

Remark: The ring $\mathbb{Z}$ of rational integers is FPF but lacks the second part of the definition of left continuous.

Summing up the above we obtain:

3.14A. THEOREM (Page [81])

A regular ring is right FPF iff R is self-injective and of bounded index.

3.14B. COROLLARY

A regular ring is left FPF iff it is right FPF.

Proof. By Utumi [63, Thm. 1.4] a strongly regular ring is left self-injective iff it is right self-injective.

The next result is a theorem of Utumi.

3.15. THEOREM (Utumi [63])

If R is a right nonsingular ring with maximal right quotient ring Q, then the f.a.e.:

3.15.1. If I is a right ideal, then ${}^{\perp}I = 0 \Rightarrow I$ is essential.

3.15.2. Every complement right ideal is an annulet.

3.15.3. If L is a nonzero left ideal of Q, then $L \cap R \neq 0$.

When these conditions hold, then R is said to be right cononsingular. Any right cononsingular ring is left nonsingular.

Proof. This follows from Utumi [63], Theorem 2.2 p.144, except for the last sentence. Hence, suppose that 3.15.1 holds, and let L be any left ideal $\neq 0$ such that $K = L^{\perp} \neq 0$. Let C be a right complement of K. Then, $J = K + C$ is an essential right ideal, and R right nonsingular implies that K is not essential, so $C \neq 0$, and also ${}^{\perp}J = {}^{\perp}K \cap {}^{\perp}C = 0$ hence $L \cap {}^{\perp}C = 0$. Now 3.15.2 implies that $C = ({}^{\perp}C)^{\perp}$, and then 3.15.1 implies that ${}^{\perp}C \neq 0$ so that L is not essential. This proves that R is left nonsingular.

A ring is (co)nonsingular if it is right and left (co)nonsingular.

THEOREM. (Utumi [63]).

A right nonsingular ring R is cononsingular iff $Q^r_{max} = Q^{\ell}_{max}$, that is, R has the same right as left maximal quotient ring.

This also appears in Stenstrom [75], p.252, Theorem 4.9. As remarked earlier, the next result holds in general for commutative FPF rings.

3.16A. THEOREM. (Page [83])

If R is nonsingular right FPF, then $Q(R)$ is both the left and right maximal quotient ring.

Proof. In order to show that $Q(R)$ is the left quotient ring of R it is enough to show that R is left essential in $Q(R)$. Let $y \varepsilon Q$. Let $e = e^2$ be such that $eQ = yQ$. Then yR is essential in $eR + yR$. Let $B = yR^{\perp}$. Then B is a closed two sided ideal, hence a direct summand. So, as above, we can assume $yR^{\perp} = 0$. Then $eR + yR$ is faithful, hence generates R. Let $f: eR + yR \to R$ with $f(e) = x = xe$ and $f(y) = z$. There is an essential right ideal W such that $yw \varepsilon eR$ for all $w \varepsilon W$. Now $xyw = zw$ for all $w \varepsilon W$, so $xy = z \varepsilon R$. Since yR is essential in $eR + yR$, $z \neq 0$.

3.16B. COROLLARY. (Page [83])

Let R be a right FPF prime ring. Then R is both left and right Goldie.

Proof. We have that $Q(R)$ is the right and left maximal ring of quotients and when R is prime, it follows that $Q(R)$ is prime and that $Q(R) = D_{n\times n}$, where D is a division ring, with n the bound on the index of nilpotency of $Q(R)$. The corollary follows.

3.16C. COROLLARY.

A semiprime right FPF ring with a.c.c. on annihilators is right and left Goldie.

Proof. This follows from 3.4 and 3.15.

3.16D. COROLLARY.

A prime regular right FPF ring R is simple Artinian.

3.17. LEMMA.

If R is a right nonsingular ring, and M an A-module where A is a subring of $\hat{R} = Q^r_{max}(R)$ containing R, then for any finitely generated right R-module M_1 contained in a right A-module M, the kernel of the canonical map $M_1 \otimes_R A \to M$ is contained in the singular A-submodule of $M_1 \otimes_R A$. Thus, if $M = x_1A + \cdots + x_nA$ is finitely generated, then M generates mod-A provided that $M_1 = x_1R + \cdots + x_nR$ generates mod-R.

Proof. The first assertion is well known, but we prove it for completeness. If $x_1, \cdots, x_n$ generate M_1, and if $a_1, \cdots, a_n \in A$ are such that $\sum_{i=1}^n x_i a_i = 0$, then there is an essential right ideal J of R such that $a_i J \subset R$, for any i, and then for $r \in J$

$$(\textstyle\sum_{i=1}^n x_i \otimes a_i)r = \sum_{i=1}^n x_i \otimes a_i r = (\sum_{i=1}^n x_i a_i r) \otimes 1 = 0$$

so that $\Sigma_{i=1}^n x_i \otimes a_i$ annihilates JA, which is an essential right ideal of A, so that the kernel K of the stated canonical map is contained in the singular A-submodule as asserted. Thus, if M_1 generates mod-R, then $M_1^n \approx R \oplus X$

in mod-R, for an integer $n > 0$. Then:

$$(M_1 \otimes A)^n \approx M_1^n \otimes A \approx A \oplus (X \otimes A)$$

that is, $M_1 \otimes A$ is a generator of mod-A; by 3.1A.2, so is M.

3.18. THEOREM. (Faith [76c])

Let R be a prime right FPF ring, and let A be a subring of $\hat{R} = Q^r_{max}(R)$ containing R. Then every finitely generated faithful right A-module $M \neq \text{sing}\, M$ is a generator of mod-A. In particular, any finitely generated right ideal $\neq 0$ generates mod-A.

Proof. Let $x_1, \ldots, x_n$ be a finite set of generators of M. If $M_1 = \Sigma_{i=1}^n x_i R$ is a faithful R-module, then by 3.17, M is a generator of mod-A. (Thus, A is right FPF provided only that M_1 is faithful for all such M.) We shall show that the existence of a nonzero ideal B of R annihilating M_1 implies that sing M = M contradicting the hypothesis. (Note that it is possible, in general, for M = sing M, M f.g. and M still to be faithful, see the proof of 3.11B).

To do this let $y = \Sigma_{i=1}^n x_i t_i$, with $t_i \in A$, be any element of M, and choose an essential right ideal J of R such that $t_i J \subset R$, $i = 1, \ldots, n$. Then, $yJ \subset M_1$, so that $yJB = 0$, that is, $y^\perp \supset JB$. Since any ideal in a prime ring is an essential right ideal, then $J \cap B$ is an essential right ideal. Now if I is a nonzero right ideal of R, then $J \cap I \neq 0$ by essentiality of J, and primeness implies that $(J \cap I)B \neq 0$. But

$$0 \neq (J \cap I)B \subset JB \cap I$$

proving that JB is an essential right ideal of R. Thus $y^\perp \cap R$ is an essential right ideal K of R, and the fact that $\hat{R}$, whence A, is a rational extension of R implies that KA is an essential right ideal of A contained in

$y^{\perp}$. Thus, $M = \text{sing } M$, contrary to the hypothesis. This proves that any finitely generated right A-module $M \neq \text{sing } M$, in particular, any nonzero right ideal of A generates mod-A.

<u>3.19</u> <u>THEOREM</u>. (loc. cit.)

<u>Let</u> R <u>be a right nonsingular prime ring such that</u> $Q = Q^r_{max}(R)$ <u>is also a left quotient ring of</u> R. <u>Further, assume that every finitely generated nonzero right ideal of</u> R <u>generates mod-R (Thus,</u> R <u>can be any right FPF, or any simple ring.)</u>. <u>Then,</u> Q <u>is a simple right</u> FP^2F <u>ring, and every right Q-module</u> $M \neq \text{sing } M$ <u>generates mod-Q</u>.

<u>Proof</u>. We start with the last assertion first. We know from 3.7A that M generates mod-Q iff M/sing M does. Since M/sing M is nonsingular, we may assume that sing M = 0. For a right nonsingular ring R and right module X with $x \in X$, since $I \cap x^{\perp} = 0 \Rightarrow xI \approx I$ we have sing $X \neq X$ iff $\text{Hom}_R(X,Q) \neq 0$. Thus, we may assume that there is a nonzero Q-module homomorphism $f:M \to Q$. (In this application, we have assumed that $R = Q$). Clearly M generates mod-Q if f(M) does, so we may assume that $M \subset Q$. So far, we have not assumed that M is finitely generated. We now do, say $M = x_1Q + \cdots + x_nQ$. Since Q is a left quotient ring, there is an essential left ideal J of R such that $Jx_i \subset R$, $i = 1,\cdots,n$, and hence, $JM_1 \subset R$, for $M_1 = \sum_{i=1}^{n} x_iR$. Since J is an essential left ideal, and since R is left nonsingular, then $JM_1 \neq 0$, so that $yM_1 \neq 0$, for some $y \in J$. This shows that there is a nonzero homomorphism $M_1 \to R$, and hence that M_1 generates mod-R, since yM_1 does. Then, by 3.18, M generates mod-Q. In particular, any principal right ideal $M \neq 0$ generates mod-Q. Now if I is any ideal $\neq 0$ in Q, then since Q is a regular ring, any principal right ideal $\neq 0$ is generated by an idempotent e, so I contains such a right ideal $M = eQ$. Since M generates mod-Q, then $\text{trace}_Q M = Q$. But if $f \in \text{Hom}_Q(M,Q)$, then $f(eQ) = f(e)eQ$ is contained in I, that is, $Q = \text{trace}_Q M \subset I$, so $I = Q$, proving that Q is

simple. Then for arbitrary $M \neq \text{sing}\, M$, there is a nonzero map $f: M/\text{sing}\, M \to Q$. By simplicity of Q every nonzero right ideal of Q generates mod-Q, hence im f, whence $M/\text{sing}\, M$, whence M, generates mod-Q. Finally, over a regular ring Q, any finitely presented module $M \neq 0$ is torsionless hence nonsingular, hence generates mod-Q. This proves that Q is right FP^2F.

3.20. PROPOSITION. (loc. cit.)

For a ring R the f.a.e.:

(1) R is prime and right FPF

(2) Every finitely generated right ideal $\neq 0$ generates mod-R, and $M^* \neq 0$ for every finitely generated faithful right module M.

(3) R is right bounded, every finitely generated right ideal $\neq 0$ generates mod-R, and $M^* \neq 0$ for any finitely generated nonsingular right module $M \neq 0$.

Proof. (1)$\Rightarrow$(3). Any prime right FPF ring R is right bounded. The other requirements for (3) follow from 3.1A and from the fact that nonzero right ideals are faithful in a prime ring.

(3) $\Rightarrow$(2). Let M be finitely generated faithful, say $M = \sum_{i=1}^{n} x_i R$. Since M is faithful, then no ideal $B \neq 0$ annihilates every x_i, $i = 1,\ldots,n$. This implies that not every x_i has essential right annihilator J_i in R, since otherwise, by right boundedness, $x_i A_i = 0$ for an ideal $A_i \neq 0$, and then $x_i B = 0$ for $B = \cap_{i=1}^{n} A_i$. Since every finitely generated right ideal $\neq 0$ generates mod-R, every such is faithful, hence R is prime, and so $B \neq 0$ annihilates every x_i, contrary to faithfulness of M. This proves that $\bar{M} = M/\text{sing}\, M \neq 0$. However, sing $\bar{M} = 0$, that is $\bar{M}$ is nonsingular, hence $\bar{M}^* \neq 0$, that is, there is a nonzero homomorphism $f: \bar{M} \to R$. Since $J = f(\bar{M})$ generates mod-R, then $\bar{M}$ generates mod-R, and hence M generates mod-R, inasmuch as a module generates mod-R whenever an epic image does. This remark also suffices for the proof of (2) $\Rightarrow$(1). (Compare 5.4A).

3.21. PROPOSITION. (Faith [77b])

Let R be a prime left FPF ring, with right and left maximal quotient ring Q.

3.21.1 *If an ideal I is a generator of* mod-R, *then I is finitely generated and projective in* mod-R.

3.21.2. *If I also generates* R-mod, *then I is invertible in the sense that*

$$II^* = I^*I = R$$

where

$$I^* = \mathrm{Hom}_R(I,R) = \{q \in Q \mid qI \subset R\}$$

Proof. Let I be an essential right ideal of R. Since $Q = \hat{R}$ is an injective R-module, any element $f \in A \approx \mathrm{End}\ I_R$ is induced by left multiplication by an element of Q, since f extends to a mapping belonging to

$$\mathrm{Hom}_R(\hat{R},\hat{R}) \approx \hat{R} = Q.$$

Moreover, if $q \in Q$ induces f, and if $f = 0$, then $qI = 0$, so $q \in \mathrm{sing}\ \hat{R}_R = 0$. This permits us to identify

$$A = \mathrm{End}\ I_R = \{q \mid qI \subset I\}$$

with a subring of Q. Then, if I is an ideal of R, essential as a right ideal, clearly $A \supset R$.

Proof of 3.21.1. Now assume that I is an ideal which generates mod-R. Then, by the Morita Theorem 1.2, I is finitely generated projective over $A = \mathrm{End}\ I_R$, and $R = \mathrm{End}_A I$ canonically. Since R is prime, then I is left faithful, so that A embeds in Q as indicated above. Write $I = Ax_1 + \cdots + Ax_n$ for generators $x_1, \cdots, x_n$ of I in A-mod, and let $I_1 = Rx_1 + \cdots + Rx_n$. By primeness of R, I_1 is a faithful left ideal of R, hence generates R-mod by the left FPF hypothesis, and then by the left - right symmetry of 3.17, we have that I generates A-mod. Then by the Morita Theorem again, I is finitely generated projective over $R = \mathrm{End}_A I$.

<u>Proof</u> of 3.21.2. We next prove that $I^*I = I\,I^* = R$ under the stated assumptions. Since Q is injective, every $f \in \mathrm{Hom}_R(I,R) = I^*$ is induced by an element $q_f \in I^* = \{q \in Q \mid qI \subset R\}$. Nonsingularity of R, and essentiality of I, imply that q_f is unique, and that the mapping $f \quad q_f$ is an embedding of R-modules. Therefore, henceforth, we write I^* in place of I'. In this notation, then I^*I represents the trace ideal of I. Since I generates mod-R, then $I^*I = R$. Moreover, since Q is also the left quotient ring of A, then $B = \mathrm{End}_A I$ can be constructed within Q as $B = \{q \in Q \mid Iq \subset I\}$, which is the same construction as $\mathrm{End}_R I$, that is $B = \mathrm{End}_A I = \mathrm{End}_R I$. However, the fact that I generates R-mod implies by the Morita Theorem 1.2 that $R = \mathrm{End}\, I_B = \mathrm{Biend}_R I$. But since A is represented as $A = \{q \in Q \mid qI \subset I\}$, and since $R = \mathrm{End}\, I_B$ canonically, then $R = A$. Then, the fact that I generates R-mod means that $II' = R$, for $I' = \{q \in Q \mid Iq \subset R\}$, and then $I^*I = R$ implies that $I^* = I'$. Therefore, I is invertible.

<u>3.22.</u> <u>COROLLARY</u>.

<u>If R is a prime FPF ring, then any ideal I of R which is finitely generated as a right ideal is projective in mod-R. Moreover, any ideal I which is a generator of mod-R is finitely generated and projective* in mod-R.</u>

<u>Proof</u>. The first statement follows from the second inasmuch as R is prime (so any nonzero ideal is an essential faithful right ideal) and right FPF (so any finitely generated nonzero right ideal is a generator). Next, FPF implies by 3.6 that R is right (and left) nonsingular, and by 3.16A, the maximal right quotient ring Q is also the maximal left quotient ring, so that 3.21.2 and its right-left symmetry applies to give the desired conclusions.

---

*Cf. The rank-1 theorem of the first author's paper [82a]: any generator of mod-R contained in the injective hull is finitely generated projective. This holds for any commutative ring R.

3.23. THEOREM. (Faith [77b])

Let R be any right nonsingular ring, let Q denote the maximal right quotient ring, and assume that Q is also a left quotient ring of R.

3.23.1. *If I is an ideal which generates R-mod, then I generates A-mod where* $A = \text{End } I_R = \{q \in Q \mid qI \subset I\}$.
3.23.2. *If I is an ideal which generates both R-mod and mod-R, then I is finitely generated projective in mod-R.*
3.23.3. *If Q is also the maximal left quotient ring of R then any ideal I which generates both R-mod and mod-R is invertible.*

Proof of 3.23.1. Since any generator is faithful, by 3.15.1 we know that I is an essential right ideal, so that $A = \text{End } I_R$ embeds in Q as stated. (See the proof of 3.21). Since I generates R-mod, then as in the proof of 3.16, $A \otimes {}_R I$ generates A-mod. However the kernel of the epimorphism $A \otimes {}_R I \to I$ is contained in the singular submodule (by virtue of the fact that Q is a left quotient ring of R, a fact which also implies that sing M as a left R-module equals sing M as a left A-module, as in the proof of 3.5). It then follows from 3.17 that I generates A-mod.

Proof of 3.23.2. By Morita Theorem 1.2, when I generates mod-R, then I is finitely generated projective over $A = \text{End } I_R$, and $R = \text{End}_A I$. Since I generates A-mod by 1., this implies that I is finitely generated projective in mod-R.

Proof of 3.23.3. The verbatim proof of 3.21.2 suffices, without assuming that R is prime, since as noted in 3.15.1, faithfulness on one side implies essentiality on the other. Thus, I is essential on both sides, and $A = \text{End } I_R$ (also $B = \text{End}_R I$) embed in Q canonically.

By Theorem 3.4, any semiprime right FPF ring with a.c.c. on ideals has a decomposition as a finite product of prime rings. The next theorem uses this in the proof.

<u>3.24</u>. <u>THEOREM</u>.

<u>A semiprime Noetherian FPF ring</u> R <u>is a finite product of prime FPF rings in which every ideal is projective on both sides.</u>

<u>Proof</u>. Any right Noetherian ring R is right Goldie, so R is a semiprime Goldie ring, and 3.23 applies. By 3.23, any ideal $I \neq 0$ which is essential or faithful on the right is projective on both sides. However, if P is any ideal, then the annihilator Q of P contains any right complement K of P, since $P \cap K = 0 \Rightarrow KP = 0$. Thus, $P + Q$ is an essential right ideal, hence is projective. Since $P \cap Q = 0$ by semiprimeness of R, then $P \oplus Q$ is projective, and hence so is P. Left projectivity follows by symmetry. That R is a finite product of prime FPF rings follows from 3.4, and, in fact, applying this first allows us to prove 3.24 employing 3.23 instead of 3.21.

This theorem implies that $\mathbb{Z}G$ is never FPF when $|G| > 1$. See 5.23. Actually, the hypotheses imply much more: <u>every onesided ideal is projective</u>. This follows since a Noetherian prime FPF ring is hereditary (Theorem 4.10).

Another point: assume something weaker than FPF; namely, assume that R is a right bounded prime ring in which every ideal is finitely generated projective on the right. Is every right ideal projective? The answer is "yes" assuming the restricted right minimum condition (see 4.18).

<u>3.25</u> <u>PROPOSITION</u> (Faith [77b])

<u>Any semiperfect semiprime right FPF ring</u> R <u>is semihereditary and is a finite ring product of full matrix rings of finite ranks over right bounded local Ore domains which are right and left VR's</u>.

<u>Proof</u>. A semiperfect ring has no infinite sets of orthogonal idempotents, hence 3.4 applies, so we may assume that R is a finite product of prime semiperfect right FPF rings. Then 2.1C implies that each prime factor has the stated structure: $n \times n$ matrices over local Ore domains which are right VR's. However, a right VR domain D is

right (and left) semihereditary inasmuch as each finitely generated right ideal $\neq 0$ is a principal right ideal, hence a projective right ideal $\approx D$. Now right FPF is a categorical, that is, Morita invariant property, so that D is right FPF, hence right bounded, along with R. Then, an application of 3.16A will show that D is left Ore. Now any right semihereditary ring S is left semihereditary provided that S_n does not contain an infinite set of orthogonal idempotents for any full matrix ring S_n over S (Small [67]); in particular this holds if S embeds in a right or left Noetherian or semiperfect ring T, for then S_n embeds in T_n, which contains no such set. Thus, D is left s.h., and hence every f.g. left ideal I is projective, hence free, since D is local, so left Ore implies I is principal, that is, D is a left VR.

ANNULETS IN PRIME FPF RINGS

Annulets in prime FPF rings are idempotent:

3.26. PROPOSITION.

Let R be a prime right FPF ring. Then:

3.26.1. *If I is any inessential right ideal, then* R/I *generates* mod-R.

3.26.2. *Any nonzero left annulet* L *generates* R-mod, *and is idempotent. Moreover,* L *is contained in no proper ideal of* R, *that is,* LR = R.

Proof. 1. Since any ideal $\neq 0$ in a prime ring R is an essential right ideal, then R/I is a faithful module, hence generates mod-R.

Proof. 2. Since R is right nonsingular by 3.6, then any essential right ideal has nonzero left annihilator. This shows that $I = L^{\perp}$ is an inessential right ideal. Moreover, R/I is a generator of mod-R by 1., hence $L = (R/I)^*$ generates R-mod, and the canonical map $(R/I)^* \otimes R/I \to R$ is a surjection h. Now the image of h is the trace ideal of R, and therefore R is generated by $\{ab \mid a \in L, b \in R\}$, that is, R = LR. This implies that $L^2 = L$.

3.27. COROLLARY.

In a prime FPF ring, any complement right ideal $K \neq 0$ generates mod-R and $RK = R$.

Proof. By 3.15, K is a right annulet, so the left-right symmetry of 3.26 applies.

We next show that nonsingular FPF rings decompose into rings with essential socle and rings with zero socle.

3.28. PROPOSITION. (Page [83])

Suppose R is right FPF and nonsingular. Then $R = R_1 \times R_2$, where R_1 has an essential right socle and R_2 has zero right socle.

Proof. Let S be the right socle of R. Then by 3.3B S is a two sided ideal of R and is essential in a two sided ideal direct summand R_1 by 3.3B. Let $R_1 \times R_2 = R$. Clearly the socle of R_2 is zero.

In the next section we show how nice things are when $D(R) = 1$. We also show how to reduce the question of whether a nonsingular FPF ring is semihereditary to the case $D(R) = 1$ and how to relate the semihereditary FPF rings to Baer rings.

In the previous chapter we looked at semiperfect rings which, when FPF had $D(R) < \infty$ (Thm. 2.1), in fact, if they were selfinjective they were Morita equivalent to one dimensional rings (the basic ring has dimension one).

3.29. PROPOSITION. (Page [83])*

Let R be a right FPF nonsingular ring. Let e be a central idempotent of $Q(R) = Q$. Then $e \in R$.

Proof. Let e be as above. Form $Qe \cap R$, a closed two sided ideal of R. So $Qe \cap R$ is a direct summand of R, i.e. $Qe \cap R = Rf$ for some central idempotent f. Now $Qf \supset Q(Qe \cap R) = Qe$ essentially, so it follows that $Qe = Qf$. But then $e = f$.

With proposition 3.29 we can reduce our study to rings with the property that Q(R) is a matrix ring over a strongly regular ring. We can in fact "get away" with even

---

*For commutative R, a more general theorem holds without assuming non-singularity: R is integrally closed in $Q_{c\ell}(R)$. (Faith [82a]).

more.

3.30. PROPOSITION. (Page [82])

If R is right FPF and nonsingular with $Q(R) = D_n$, where D is strongly regular, then Q is a flat epimorphism* of R. If also w.gl.dim $R \leq 1$, then R is Morita equivalent to a ring S with $Q(S) = D$.

Proof. The first part follows from 3.12. Let e be the matrix unit in $Q = D_n$ with a one in the upper left corner. Let P be $eQ \cap R$. Since eQ is faithful as a Q-module and $eQ \cap R$ is an essential R submodule of eQ, it follows that $eQ \cap R$ is faithful as an R-module. Also $eQ \cap R = eR \cap R$. Now let $S = \text{End}\ eR$. Since eR is a faithful generator of R and $e^{\perp}$ is a closed right ideal of R, we claim eR is projective. This follows because eQ is flat since it is a direct summand of Q. Hence eR is flat, when w.gl. dim $R \leq 1$. Now $eR \otimes Q \approx eQ$ is Q projective so by Sandomierski [68 Thm. 2.8, p. 228], Re is projective. So R is Morita equivalent to S. Now, $\text{Hom}_R(eR, eR)$ embeds, by extension, into $\text{Hom}_{Q(R)}(eQ, eQ) = eQe = D$. We claim $Q(S) = D$. Since D is a strongly regular self-injective ring, if S is S-essential in D then it follows that $Q(S) = D$. We have that $S \approx eRe$ canonically, so if $eqe \in D$, let H be an essential right ideal of R so that $eqeH \subset R$. Then $eqeH \neq 0$ since R is semi-prime, so $eqeHe \neq 0$ and $eqeHe \subset S$. Therefore we have $Q(S) = D$.

Let R be a right FPF nonsingular ring. It is not known if R is left or right semihereditary.** The next theorem reduces the question to rings with strongly regular quotient rings. Lenzing [70] calls a ring a B-ring if R is a Baer ring (i.e. the annihilators of subsets of R are generated by idempotents) and for all integers n, the $n \times n$ matrix ring over R is a Baer ring. Proposition 3.29 shows all right FPF nonsingular rings with strongly regular quotient rings are Baer rings for any annihilator right ideal H is closed and hence $HQ = eQ$ for some idempotent and so $H = eR$.

---

*For commutative R, this follows from Faith [82a] without assuming R is nonsingular! We conjecture Q is always a flat epi of a right FPF ring R. See Open Problems.

**For commutative R, this holds by a theorem of Faith [79a] and for Noetherian semiprime R this holds by a theorem in Chapter 4.

3.31.1. THEOREM. (Page [83])

Let R be a right FPF nonsingular ring. Then the following are equivalent:

i) w.gl.dim R ≤ 1 and R is right coherent.

ii) R is left semihereditary and right coherent.

iii) R is a B-ring.

iv) R is right semihereditary.

v) For every finitely generated projective R-module P, if $\{G_\alpha\}_{\alpha\in A}$ is a collection of direct summands of P, then $\bigcap_{\alpha\in A} G_\alpha$ is a direct summand of P.

Proof. The equivalence of iii) and v) is due to Lenzing [Satz 11,70]. That ii)→i) is standard. To prove i)→ii) let I be any finitely generated left ideal of R. We know Q is right flat over R by 3.12. It follows that $Q \otimes_R I$, is a finitely generated left Q-module. So QI is a direct summand since Q is regular. This says QI is projective as a Q-module and hence I is projective as an R-module by Sandomierski [68].

In order to show i)→iii) by Lenzing [Satz 11, 70] it is enough to show each torsionless (i.e. embeddable in a product of copies of R) finitely generated left R-module is projective. But since R is right coherent, if we assume ii), every finitely generated torsionless left R-module is flat and also nonsingular. As in the proof of i) implies ii) we see that finitely generated torsionless left R-modules are projective and iii) follows. Lenzing [Satz 11,70] gives iii)→iv). That iv) implies i) is standard.

3.31.1A. COROLLARY.

Let R be a nonsingular right FPF ring. R is right semihereditary iff Q(R) is left flat and R is left semihereditary.

Proof. We established that Q(R) is the left quotient ring of R in theorem 3.16A. If Q(R) is left flat then by [theorem 5.18 Goodearl, 76] R is right semihereditary. The other implication also follows from theorem 5.18 of Goodearl [76].

3.31.1B COROLLARY.

If R is a right FPF nonsingular ring, then R is right semihereditary iff R is Morita equivalent to a ring S which is right semihereditary and $Q(S)$ is strongly regular.

Proof. If R is right semihereditary and right FPF corollary 3.31.1A tells us that $Q(R)$ is left flat and R is left semihereditary. Theorem 5.18 of Goodearl [76] implies all finitely generated nonsingular right R-modules are projective. Let e be an abelian idempotent of $Q(R)$ such that eR is faithful. Such an idempotent exists since $Q(R)$ is a product of matrix rings over strongly regular rings and is of bounded index. So eR is a projective generator and $S = \mathrm{End}_R(eR)$ is Morita equivalent to R. Now e can be chosen so that $e = \pi e_\alpha$ where $Q(R) = \Pi_{\alpha \varepsilon A} R_\alpha$, R_α a matrix ring over D_α, where D_α is strongly regular, and e_α is the matrix with one in the upper left corner. It follows, as in proposition 3.30 that $Q(S) = \Pi_{\alpha \varepsilon A} D_\alpha$ and the result follows.

3.31.1C COROLLARY

If R is FPF on both sides and nonsingular, then R is right semihereditary iff left semihereditary.

Proof. By 3.12 $Q(R)$ is flat on both sides and use 3.31.1A.

3.32. PROPOSITION

Let R be a non-singular right FPF ring with $Q(R)$ strongly regular. Then each right ideal H of R contains a nonzero two sided ideal which is essential as a right ideal in H.

Proof. Let $r \varepsilon R$, be such that rR is essential in eR with e a central idempotent. In case $(eR/rR \oplus (1-e)R)$ has zero right annihilator it must generate mod-R, and hence eR/rR generates eR, which is impossible. This implies that $(eR/rR)^{\perp} \cap eR = W \neq 0$. But since e is central, W is a two sided ideal contained in rR. One easily sees W is the largest ideal contained in rR. If

$W \cap B = 0$, where B is a right ideal contained in rR, then, if B is not zero, it must contain a nonzero ideal, by the above. The maximality of W implies B is zero. It is an easy matter to generalize the previous statement to an arbitrary right ideal.

A ring is called <u>right fully idempotent</u> if $H^2 = H$ for every right ideal H.

As a consequence of theorem 3.1A we have:

<u>3.33</u>. <u>PROPOSITION</u>. (Page [83])

<u>Let</u> R <u>be a right fully idempotent right FPF ring</u>. <u>Then for each</u> $a \varepsilon R$ <u>the ideal</u> RaR <u>is an ideal direct summand of</u> R.

<u>Proof</u>. We claim that RaR is closed. Let B be maximal so that $RaR \cap B = 0$. By semi-primeness B is two sided and is closed since it is a complement. Hence $B = eR$, e a central idempotent. Now $aR \oplus B$ is faithful and finitely generated so generates R. Because $aRaR = aR$ and $(aR)^{\perp} = B = {}^{\perp}(RaR)$ we have that $trace(aR \oplus B) = R$ and $trace(aR) = RaR$. So $B \oplus RaR = R$.

<u>3.34</u> <u>PROPOSITION</u>

<u>Let</u> R <u>be a right fully idempotent right FPF ring with</u> Q(R) <u>strongly regular</u>. <u>Let</u> M <u>be a maximal right ideal</u>. <u>Then</u> M <u>is two-sided or</u> $M \supset E$, E <u>a finitely generated essential right ideal</u>.

<u>Proof</u>. Let M be a maximal right ideal and $S = R/M$. We claim that either every element of S is annihilated by a regular element of R or, no element of S is annihilated by a regular element of R. To see this suppose $s^{\perp}$ contains no regular elements. Let $y \varepsilon s^{\perp}$. Now $Q(R) = Q$ is strongly regular so there exists $q \varepsilon Q$ such that $yq = e$ is a central idempotent in R by proposition 3.29. Next let $H = S^{\perp}$, the largest two-sided ideal contained in M. Either $e \varepsilon M$ or $(1-e) \varepsilon M$ because M is a maximal right ideal. So either $(1-e) \varepsilon H$ or $e \varepsilon H$. If $(1-e) \varepsilon H$, then $ey + (1-e)$ is regular because in Q(R),
$(ey + (1-e))(eq + (1-e)) = yqe + (1-e) = e + 1-e = 1$. Also,

$Se = S$ if $Se \neq 0$ so $S(ey + (1-e)) = 0$, contradicting the fact that $s^{\perp}$ contained no regular elements. But if $e \varepsilon H$, then $y \varepsilon H$. So $s^{\perp} = H$. This means $S \approx R/H$ and hence that $M = H$, and is two sided. If $s^{\perp}$ contains a regular element for every $s \varepsilon S$, then $M = (\bar{I})^{\perp}$ must contain a regular element, so $MQ = Q$, since regular elements in Q are units. In this case let E be the right ideal generated by one of these regular elements in M.

3.35. PROPOSITION

Let R *be a fully idempotent prime right* FPF *ring with* $Q(R)$ *strongly regular. Then* R *is a division ring.*

Proof. Let $a \varepsilon R$. We know aR is an essential submodule of eR with e central, i.e. aR is an essential left ideal. Now RaR is a summand, so $RaR = R$. We have that $(R/aR)^{\perp} \neq 0$ because it is singular. So aR contains a nonzero ideal which is generated by idempotents and is essential as a right ideal. We have $aR = R$ since R is prime and idempotents are central.

3.36. THEOREM

Let R *be a right fully idempotent right* FPF *ring with* $Q(R)$ *strongly regular. If for each right essential two sided ideal* I, R/I *has d.c.c. on principal ideals, then* R *is regular.*

Proof. Let $a \varepsilon R$, as in the previous proof we can reduce to the case $RaR = R$. This implies a is a regular element of R since aQ is a finitely generated ideal of Q and is therefore Q. Now aR contains a two sided essential ideal, I, since $(R/aR)^{\perp} \neq 0$. But if $a^n \varepsilon I$ for some n, a^nR is a generator, hence $I = R$ as we have seen. So $a^n \notin I$ for all n. It follows that $a^nR = a^{n+1}R$ for some n by the d.c.c. on principal ideals of R/I. But this says $a^n = a^{n+1}r$ for some $r \varepsilon R$, and $a^n(1-ar) = 0$. Since a^n is regular, $1 - ar = 0$ and $aR = R$ from which the result follows.

3.37. PROPOSITION

If R is a simple right FPF ring, then R is simple Artinian.

Proof. This follows from Proposition 2.8.

3.38 THEOREM (Page [82])

If R is a right fully idempotent right FPF left (or right) Goldie ring with $Q(R)$ strongly regular, then R is semisimple Artinian.

Proof. Let M be a maximal two sided ideal. If M is essential as a right ideal M contains a regular element, r, since R is semiprime Goldie by Theorem 3.16A and theorem 3.16C. By proposition 3.33 $RrR = R$. So M is not essential. Since M is maximal and not essential $M^{\perp} \neq 0$. It follows by the semiprimeness of R that $R = M \oplus M^{\perp}$ and R/M is a simple right FPF ring. Since M is a ring direct summand of R we can repeat the above inductively to obtain $R = R_1 \times R_2 \times \cdots \times R_n$, where each R_i is a simple FPF ring. Using proposition 3.37 on each R_i completes the proof.

For nonsingular rings with esential right socle we have:

3.39 PROPOSITION (Page [82])

If R is a right FPF nonsingular ring with esential right socle, then R is a subdirect product of semisimple Artinian rings, $\{S_i\}_{i \in I}$, with a bound on Goldie dimension of the S_i, $i \in I$.

Proof. Let S be the socle of R. Because R has no infinite sets of independent isomorphic right ideals, as remarked just before definition 4, we can write $S = \sum_{\substack{i \in I \\ j \in n_i}} S_{ij}$ where $S_{ij} \approx S_{i1}$ for all j and k, $S_{ij} \not\approx S_{ik}$ if $i \neq \ell$, and there is an integer N such that $n_i \leq N$

for all $i \in I$, the bound coming from the bound on the index of nilpotence of $Q(R)$. Let $S_i = \sum_{j \le n_i} S_{ij}$. Each S_i is a two sided ideal of R, hence essential in a two sided ideal A_i of R which is a two sided direct summand of R. Now A_i is a right FPF nonsingular ring with S_i as essential right socle. In A_i, ${}^{\perp}S_i = 0$, and S_i is finitely generated so S_i generates A_i. It follows that A_i is semisimple and is in fact S_i. So $S_i = e_i R$ where e_i is a central idempotent. Let T be the product of the S_i, $i \in I$. Then the map $r \to \pi\{e_i r\}$ gives the required embedding.

For V-rings we have:

3.39A. COROLLARY

If R is a right FPF, right V-ring with essential right socle, then there exists semisimple Artinian rings $\{S_i\}_{i \in I}$, with a bound on the Goldie dimension of the S_i's, such that $Q(R) = \prod_{i \in I} S_i \supset R \supset \sum_{i \in I} S_i$.

Proof. When R is a V-ring each of the S_i in proposition 3.37 is injective, and since the idempotents $\{e_i\}_{i \in I}$ of proposition 3.37 are central it follows that R is essential in $\prod_{i \in I} S_i$ so that $\prod_{i \in I} S_i = Q(R)$.

4 GOLDIE PRIME FPF RINGS WITH RRM AND THE STRUCTURE OF NOETHERIAN PRIME FPF RINGS

In this chapter bounded Dedekind rings are characterized as the FPF rings among Noetherian prime rings (Theorems 4.6 and 4.15), employing the results of Chapter 3 on prime FPF rings, and the Asano-Michler theorem presently stated. As a corollary, we obtain that every Noetherian FPF prime ring is CFPF (Theorem 4.10), so every such ring is fully bounded and fully Goldie (in the sense that every factor ring has the stated property).

A prime Goldie ring R is right FPF iff R is right bounded and every nonzero finitely generated right ideal generates mod-R (4.7). In a right bounded Goldie prime ring, this happens whenever for any finitely generated right ideal $J \neq 0$ the ideal RJ generates mod-R (4.12), in particular if every nonzero ideal generates mod-R. Thus, a Noetherian prime ring R is FPF iff it is bounded and every ideal $\neq 0$ is a generator on both sides (Theorem 4.15).

Other theorems in this chapter aim at a classification of prime FPF rings which are not assumed to be Noetherian, that is, which <u>may</u> not be Dedekind. However, a right bounded right Goldie prime ring R with the right restricted minimum condition (RRM) is right Noetherian (Lemma 4.19A), hence any prime FPF ring R with RRM is right Noetherian and right hereditary. Such a ring is conjecturally a Dedekind prime ring. For rings with left restricted minimum condition, this conjecture is verified by Theorem 4.20.

A ring R is said to satisfy the <u>restricted right minimum condition</u> (RRM) if for every essential right ideal, R/I is a right Artinian module, and similarly for RLM.

4.1 DEFINITION

A ring R is a Dedekind (prime) ring if it has any one of the following equivalent properties:

DP1. R is a hereditary Noetherian prime (HNP) ring with no nontrivial idempotent ideals.

DP2. R s an HNP which is a maximal order in its quotient ring Q. (Thus, a ring S satisfying $Q \supset S \supset R$, and $aSb \subset R$ for regular elements a, $b \in Q \Rightarrow S = R$.)

DP3. R is an HNP ring in which every ideal is invertible.

Cf. Robson [68, Theorem 2.1] for DP3 and Eisenbud-Robson [70b, Theorem 1.2] for DP1. Actually, if R is bounded, the "H" hypothesis (in HNP) in DP3 is superfluous.

4.2A THEOREM (Asano-Michler)

A Noetherian bounded prime ring in which every ideal $\neq 0$ is invertible is a Dedekind ring.

By the theorem of Asano and Michler; the hypotheses imply that R is hereditary (see Griffith and Robson [70] for a short proof and references). Lenagan [71] gave an elementary proof of the following generalization:

4.2B THEOREM (Lenagan [71]

A right Noetherian right bounded prime ring R in which every ideal $\neq 0$ is invertible is a bounded Dedekind ring.

Asano [50] assumed that R satisfied RRM and RLM, which implies that R is Noetherian; Michler [69] eliminated the RRM and RLM but retained the Noetherian hypothesis, and Lenagan gave a much simpler proof of Michler's theorem while only requiring right Noetherian.

Before giving the proof of 4.2B we need an observation:

4.2C PROPOSITION

Let R be a ring which is a subring of a ring Q. Furthermore, suppose that every ideal $I \neq 0$ in R is invertible in Q, namely, $I^{-1}I = II' = R$, where

$I^{-1} = \{q\varepsilon Q \mid Iq \subset R\}$ <u>and</u> $I' = \{q\varepsilon Q \mid Iq \subset R\}$. <u>Then</u>:

(i) <u>every prime ideal</u> $\neq 0$ <u>in</u> R <u>is maximal</u>;

(ii) <u>any two maximal ideals commute</u>;

(iii) <u>any ideal</u> $I \neq 0$ <u>is a product of maximal ideals</u>,

$I = P_1^{e_1} \cdots P_t^{e_t}$, <u>where</u> P_i <u>is an integer, and</u> P_i <u>is maximal</u>, $i = 1,\ldots,t$.

(iv) <u>The factorization</u> (iii) <u>is unique up to order</u>.

<u>Proof</u>. Clearly $I^{-1}I = II' = R \Rightarrow I^{-1} = I'$. Moreover, $II^{-1} = R \Rightarrow I$ is finitely generated as a right R-module, since $I = \Sigma_{i=1}^{n} x_i q_i$ for some $x_i \varepsilon I$, $q_i \varepsilon I = I^{-1}$, $i = 1,\ldots,n$, hence $x = \Sigma_{i=1}^{n} x_i q_i(x)$ $\forall x \varepsilon I$, so I is f.g. and projective on the right. Similarly I is f.g. projective on the left.

Thus, R has a.c.c. on ideals, and hence, every ideal $\neq 0$ is a product of prime ideals. (A maximal counterexample, I leads to the familiar contradiction: I is not prime, hence $I \subset P$ maximal; but $P^{-1}I \supset I \Rightarrow P^{-1}I = Q$ is a product of maximal ideals, and hence, then so is $I = PQ$.) This implies (i) and (iii); (ii) is an exercise; and then (iv) is trivial.

<u>4.3A DEFINITION</u>

<u>Let</u> R_1 <u>and</u> R_2 <u>be right orders in a ring</u> Q. <u>Then</u>

(i) <u>they are equivalent</u> ($R_1 \overset{Q}{\sim} R_2$) <u>if there exist regular elements</u> a_1, b_1, a_2, b_2 <u>of</u> Q <u>such that</u> $a_1 R_1 b_1 \subset R_2$ <u>and</u> $a_2 R_2 b_2 \subset R_1$;

(ii) <u>they are right equivalent</u>, $R_1 \overset{r}{\sim} R_2$, <u>if there exist regular elements</u> a_1, a_2 <u>of</u> Q <u>such that</u> $a_1 R_1 \subset R_2$ <u>and</u> $a_2 R_2 \subset R_1$;

(iii) <u>they are left equivalent</u>, $R_1 \overset{\ell}{\sim} R_2$, <u>if there exist regular elements</u> a_1, a_2 <u>of</u> Q <u>such that</u> $R_1 a_1 \subset R_2$ and $R_2 a_2 \subset R_1$.

4.3B DEFINITION

A right order R in a ring Q is a maximal $\underset{\sim}{Q}$ right order if it is maximal amongst the $\underset{\sim}{Q}$ equivalent right orders of Q. Maximal $\underset{\sim}{r}$ right order and maximal $\underset{\sim}{\ell}$ right order are defined likewise.

It is known [ARMC, 10.25] that R is a maximal $\underset{\sim}{Q}$ right order iff R is a maximal $\underset{\sim}{r}$ right order and a maximal $\underset{\sim}{\ell}$ right order. Also, the Dedekind rings are precisely the maximal $\underset{\sim}{Q}$ orders if Q is simple Artinian, (Robson [68]).

Proof. (of 4.2B, Lenagan). We first prove for all prime ideals $P \subset R$, that R/P is semisimple (Artinian). By 4.2C, P is maximal. Since R is right Noetherian, let $\bar{R} = R/P$ and by Goldie's theorem, $\bar{R}$ has a semi-simple right quotient ring $Q(\bar{R})$, so it suffices to prove that $\bar{R} = Q(\bar{R})$; and for this it suffices to prove for all regular $\bar{c} \varepsilon \bar{R}$, that $\bar{c}^{-1} \varepsilon \bar{R}$. Since $\bar{R}$ is right Noetherian, it suffices to prove $\bar{c}\bar{R} = \bar{R}$.

We first prove that c is regular in R, and since R is right Goldie we need only prove $c^{\perp} = 0$. Let $y \varepsilon c^{\perp}$. Then $y \varepsilon P$, and so $yP^{-1} \subset R$, and $cyP^{-1} = 0$, so $yP^{-1} \subset P$, that is, $y \varepsilon P^2$. By induction, $y \varepsilon Y = \cap_{n=1}^{\infty} P^n$. First suppose $Y \neq 0$. Then, $Y = P_1^{e_1} \ldots P_t^{e_t}$ as in (iii) of 4.2C, say $Y = PQ$, where $P = P_1$, and $Q = P^{-1}Y$. Since $PQ \subset P^n$ $\forall n$, then $Q \subset P^n$ $\forall n$, so $Q \subset Y \subset Q$, that is, $Y = PY$, so $P = R$, contrary to the assumption on P. Thus, $Y \neq 0$, so $y = 0$, and $c^{\perp} = 0$ as asserted.

Since c is regular, by the Goldie theorems cR is essential, so R right bounded and there is an ideal $B \neq 0$ contained in cR. Choose B to be the largest ideal $\subset cR$. If $B \not\subset P$, then $B + P = R$, hence $CR + P = R$, so $c\bar{R} = \bar{R}$ as desired.

The second alternative $B \subset P$ cannot occur. For then, $B \subset cR$ which implies $B \subset cP$, which implies B^{-1} cR which implies $B = BP^{-1}$ by the choice of B. But then $P^{-1} = R$ and $P = R$, contrary to the choice.

In order to prove that R is right hereditary, it suffices to prove that any essential right ideal I is projective. Then, I contains an ideal $B \neq 0$, and B is a product $P_1^{e_1} \ldots P_t^{e_t}$ of maximal ideals $P_1, \ldots, P_t$. By the Chinese remainder theorem,

$$R/B \cong \Pi_{i=1}^{t} R/P_i^{e_i}$$

is a product of completely primary Artinian rings $R/P_i^{e_i}$, $i = 1, \ldots, t$. This proves that R satisfies the right and left restricted minimum conditions, and, in particular, R/I has a composition chain of finite length. First suppose that R/I has length 1. Then, the annihilator ideal P is maximal, so R/P is semisimple, and hence $I + K = R$, with $I \cap K = M$ for some right ideal K. Since R is projective the canonical map $I \oplus K \longrightarrow I + K = R$ splits, with kernel $\approx M$, so $I \oplus K \approx R \oplus M$. Since M is invertible, M is projective, so I (also K) is projective. The general case follows by induction on n = length (R/I). Thus, if $I' \supset I$ and $V = I'/I$ is simple, we may assume that I' is projective and $V = R/J$ for a projective right ideal J. By Schanuel's lemma (ARMC, p. 436, Lemma 11.2), $R \oplus I \approx I' \oplus J$, so that I is also projective.

To obtain the fact that R is left Noetherian by Small [67], R is left semihereditary. Now if $I_1 \subset I_2 \subset I_3 \ldots$ is a chain of finitely generated left ideals, we can choose a finitely generated left ideal K so that for some n, $J_m = L_m \oplus K$ is essential for all $m \geq n$. Robson [68] shows R is a maximal right order. As in Jacobson [43, p. 121] R is a left bounded, left order in Q, so there is an ideal T contained in J_n. Now $T^{-1} \supset J_n' \supset J_{n+1}' \ldots$ so $R \supset J'T \supset J_{n+1}'T \supset \ldots \supset T$ and so $J_m' = J_{m+\ell}'$ for some large enough m and all $\ell \geq 0$. It follows that $J_m = J_{m+\ell}$ and we have a.c.c. on f.g. left ideals so that R is left Noetherian and left hereditary.

4.4 THEOREM

If R is a right Noetherian prime right FPF ring, R is a maximal $\underset{\sim}{r}$ right order in Q.

Proof. If $R \subset S \subset Q$ and $dS \subset R$ where d is a regular element of Q, then dS is a finitely generated right ideal of R. Now dS is faithful since R is prime. So by 1.10 $\mathrm{Biend}_R(dS) = R$. On the other hand, $S \subset \mathrm{Biend}_R(dS)$ so $S = R$.

4.5 PROPOSITION

Let R be a left Noetherian ring and I an ideal which is finitely generated as a left ideal. Then $\mathrm{End}_R(I_R)$ is a finitely generated left R-module.

Proof. Let $I = \Sigma_{i=1}^{n} Rx_i$. If f is in $\mathrm{End}_R(I_R)$, let $f(x_i) = \Sigma_{j=1}^{n} r_{ij}x_j$. Form the matrix (r_{ij}). This gives a left R-module embedding of $\mathrm{End}_R(I_R)$ into the $n \times n$ matrices over R. Since R is left Noetherian the result follows.

We now combine the above two results.

4.6 THEOREM.

For a ring R, the following are equivalent:

4.6.1 R is a Noetherian prime right FPF ring.

4.6.2 R is a bounded Dedekind prime ring.

4.6.3 R is a Noetherian prime left FPF ring.

Proof. Assume 4.6.1. By 3.16A we know R is Goldie on both sides. Now each two sided ideal, I, is faithful and finitely generated as a right ideal hence is a generator as a right ideal i.e. the right trace of a two sided ideal is all of R. So as in 4.2C, for a two sided ideal I, $I^{-1}I = R$. We need to show $II' = R$. Now I' is the dual of I as a left ideal and II' is the trace of I as a left ideal. If we can show I is projective as a left ideal we will have $II' = P$ and $P^2 = P$. If $P^2 = P$, then $P = R$, for $P^{-1}P = R$. Let $S = \mathrm{End}_R(I_R)$. Then $S \supset R$ and by 4.4 and 4.5 $S = R$. By 1.10 ${}_RI$ is projective over R and we have established our claim. Since R acts like a unit when multiplying ideals it follows that $I^{-1} = I'$ and by 4.1 R is Dedekind and by 1.13B R is right bounded and by 4.2 left bounded.

To prove the converse we introduce a concept (TF) which will be useful in the description of prime FPF rings.

A right R-module M is *torsion free* if no element of M other than zero annihilates a regular element of R. Any submodule of a free module is torsion free. A ring is *(right) TF* if every finitely generated torsion free (right) module embeds in a free module (Levy [63].) Note, if $R = Q(S)$ is its own right quotient ring, *every* module is torsion free. (In this case, by a result in Faith-Walker [67, p.217, Corollary 5.10], R is TF iff QF.)

Levy proved 2 main theorems on TF rings.

LEVY'S FIRST THEOREM

Let R *have right quotient ring* Q(R). *Then:*

$$R \text{ right TF} \Rightarrow Q(R) \text{ right TF}.$$

If Q(R) *is also the left quotient ring of* R, *then conversely*,

$$Q(R) \text{ right TF} \Rightarrow R \text{ right TF}$$

For a proof, see Levy [63, p.144, 5.1-2]. As noted, when Q is the two sided quotient ring, then R (and also Q) is two sided TF iff Q is QF.

LEVY'S SECOND THEOREM

If R *is semiprime right Goldie, and if* R *is right TF, then* Q(R) *is semisimple, and the left quotient ring of* R.

We shall apply Levy's theorem to characterize prime FPF rings.

4.7 THEOREM.

A ring R *is a prime right FPF ring iff* R *is a right bounded Goldie (both sides) ring in which every nonzero*

<u>finitely generated right ideal is a generator</u>.

<u>Proof</u>. Any prime right FPF ring R is right bounded by 1.13B, and a prime right FPF ring is Goldie by 3.16B. Since any right ideal is faithful in a prime ring, we see the necessity of these conditions.

Conversely, let M be any finitely generated faithful right module over a ring R satisfying these conditions. Now, a finitely generated faithful module $M = \Sigma_{i=1}^{n} x_i R$ cannot be a torsion module, since by right boundedness of R, we have $x_i^{\perp}$ contains an ideal $A_i \neq 0$, $i = 1,\ldots,n$, and then $B = \cap_{i=1}^{n} A_i$ annihilates M, a contradiction. Thus, $\bar{M} = M/t(M) \neq 0$. Since $\bar{M}$ is torsion free, and since R is TF by Levy's theorem, we have that $\bar{M}$ embeds in a free R-module. Thus, $\bar{M}^* \neq 0$, so some homomorph f(M) is a nonzero right ideal, and f(M), whence M is a generator.

Since Dedekind rings are clearly Goldie on both sides the remaining part of the proof of 4.6 is just a special case of the above.

<u>4.8</u> <u>COROLLARY</u>

<u>For a ring</u> R, <u>the following are equivalent</u>:

4.8.1 R <u>is a prime Noetherian FPF ring</u>;

4.8.2 R <u>is a bounded Dedekind ring</u>.

The proof of the next theorem requires an observation made by Michler [69].

<u>4.9</u> <u>PROPOSITION</u>

<u>If</u> A <u>is an ideal</u> $\neq 0$ <u>in a Dedekind prime ring</u> R, <u>then</u> R/A <u>is an Artinian principal ideal ring, hence is uniserial, and</u> QF.

The proof is trivial: write $A = P_1^{e_1} \ldots P_n^{e_n}$ as a product of maximal ideals, where $P_1,\ldots,P_n$ are different maximal ideals. Then, by the Chinese Remainder Theorem, $R/A \approx \Pi_{i=1}^{n} R/P^{e_i}$ a product of Artinian primary rings

$R_i = R/P_i^{e_i}$, and such that there are no ideals between P_i^j and P_i^{j+1}, for any j, $i = 1,\ldots,n$. Thus, R_i is a full matrix ring over a uniserial local ring D_i, $i = 1,\ldots,n$, and so R/A is uniserial, hence QF.

Actually, that R/A is an Artinian principal ideal ring, and that this is equivalent to being an Artinian uniserial ring is a theorem of Asano [49, 50]; Michler made the observation that R/A is Artinian self-injective; but this is obvious.

4.10 THEOREM.

4.10.1 *A right bounded Dedekind ring is left and right CFPF.*

4.10.2 *A left bounded Dedekind ring is right CFPF.*

4.10.3 *A right bounded Dedekind ring is left bounded and vice-versa.*

Proof. This follows from 4.6 and 4.9.

4.11 REMARK.

Theorem 4.6 shows that any Noetherian prime right FPF ring is HNP, but the converse is false since not every HNP is Dedekind or bounded. For example any simple principal ideal domain, e.g. the ring $k[x,\delta]$ of differential polynomials in a differential δ of a field k of characteristic 0 is a right and left PID but not bounded. In fact any simple Noetherian V ring which is not regular cannot be an FPF ring.

Also, another PID which is not FPF: for any (noncommutative) field D which is transcendental over its center, the polynomial ring $D[x]$ is a primitive ring, so is not bounded.

A ring R is right *pre-Prüfer* provided that R is a Goldie prime ring in which every finitely generated ideal $\neq 0$ generates mod-R.

We give another application of theorem 4.7.

4.12 THEOREM

A right bounded right pre-Prüfer ring is right FPF.

Proof. Let J be any finitely generated right ideal $\neq 0$. Since RJ is an epic image of a direct sum of copies of J, the fact that RJ generates mod-R implies that J does, too. (It would be interesting to know when the converse of this holds. It does, for example, when R is simple, or right PF!) Then, R is right FPF by 4.7.

4.13 THEOREM

A Goldie prime ring in which finitely generated ideals are principal right ideals is right pre-Prüfer, and hence, if right bounded, right FPF.

Proof. If J is a finitely generated ideal $\neq 0$, then $J = tR$ is a principal right ideal. Since ideals $\neq 0$ in prime rings are essential, then J contains a regular element by the theorems of Goldie [58, 60], and then t is easily seen to be regular. Thus, $J = tR \approx R$ canonically, so J generates mod-R, and hence R is right pre-Prüfer. Therefore, right bounded implies right FPF by 4.12.

A ring R is a Prüfer ring if R is a Goldie prime ring in which every finitely generated ideal $\neq 0$ is invertible. By 4.13, any prime Goldie ring in which finitely generated ideals are principal right, and principal left, ideals is Prüfer.

4.14A PROPOSITION

Let R be a bounded nonsingular ring with maximal right quotient ring also the maximal left quotient ring. If every finitely generated ideal $\neq 0$ is a generator on both sides, then R is a Prüfer ring, hence FPF.

Proof. By 4.2C every ideal which is a generator on both sides (e.g. any finitely generated ideal $\neq 0$) is invertible, and hence all that remains is to prove that R is Goldie, since R is manifestly prime. However, since finitely generated right (left) ideals are generators (see the proof of 4.12), apply 3.19, and its left-right symmetry, to obtain that R is Goldie. Moreover, R is FPF by 4.12.

4.14B COROLLARY

Any bounded pre-Prüfer ring is a Prüfer ring.

Proof. Apply 4.14A.

A ring R is pre-Dedekind provided that it is a prime Goldie ring in which every ideal $\neq 0$ is a generator on both sides. Thus, a pre-Dedekind ring is pre-Prüfer.

We now summarize one of the main aspects of Noetherian Prime FPF rings.

4.15 THEOREM

For a ring R, the following conditions are equivalent.

4.15.1 R is a Noetherian prime right (left) FPF ring.

4.15.2 R is a bounded Dedekind prime ring.

4.15.3 R is a right (left) bounded Noetherian pre-Dedekind ring.

4.15.4 R is a Noetherian prime right (left) CFPF ring.

Proof. The equivalence of 1,2, and 4 has been established by 4.6 and 4.10. Moreover, $2 \Rightarrow 3$ is trivial, while $3 \Rightarrow 1$ follows from 4.14 (also from 2.23), using the fact that R is Noetherian.

4.16 COROLLARY

A right (left) bounded Noetherian pre-Dedekind ring is (Dedekind, and hence) hereditary.

Proof. R is Dedekind by 4.15.3, hence HNP by 4.1.

4.17 THEOREM

A ring R is a finite product of bounded Dedekind prime rings iff R is a semiprime Noetherian FPF ring. When this is so, then R is CFPF.

Proof. We already have shown (see, for example, the proof of 3.4.2), that a finite product of rings is (C)FPF iff each of the factors is (C)FPF, and also 4.10 establishes the theorem for a prime ring, hence it suffices to prove that any semiprime Noetherian FPF ring is a finite product of prime rings, but this is 3.4.1.

By Definition 4.1, these rings are maximal orders in semisimple rings, so the theorem contains that Endo [67] for orders in semisimple finite dimensional algebras.

COHEN RINGS AND THE RESTRICTED MINIMUM CONDITIONS

In this section we present some applications of the results of Chapter 3 on prime FPF rings to RRM rings, and Cohen rings defined presently.

The _restricted right minimum condition_ (RRM) states that R/I is an Artinian module, for any essential right ideal I. The theorem of Webber [70] - Chatters [71], states that any hereditary Noetherian prime ring is both RRM and the left-right symmetry RLM. A ring is a _Cohen_ ring if R/P is right artinian for any prime $P \neq 0$. Any prime RRM ring is a Cohen ring with a.c.c. on ideals. Moreover, any right bounded Cohen ring in which ideals are finitely generated on the right is RRM. (In this connection, note that any Cohen ring in which ideals are finitely generated right ideals is either prime or right Artinian.) These results are theorems of Cohen and Ornstein, or modifications thereof made in Faith [75].

_4.18_ _PROPOSITION_ (Zaks [71]).

Let R _be a right bounded Cohen ring_. _Then, every ideal is finitely generated projective on the right iff_ R _is right Noetherian and right hereditary_.

For another proof, and a generalization, see Faith, _loc.cit_ (The proofs are similar to the proof of 4.2B.)

_4.18A_ _COROLLARY_

Let R _be a right bounded nonsingular Cohen ring with maximal quotient ring_ Q _which is also a maximal left quotient ring_. _Assume furthermore, that any ideal_ $\neq 0$ _is a generator on both sides_. _Then_, R _is Noetherian and_ hereditary and left bounded.

Proof. By 3.23.2, every ideal is finitely generated and projective on the right, so that 4.18 applies to yield the conclusion that R is right Noetherian and right hereditary. To see that R is left Noetherian and left hereditary the arguments of Lenagan (i.e. proof of 4.2B) apply. (Another proof: the fact that ideals are finitely generated, together with Cohen, implies that R is Noetherian, so the last part of the proof of 4.2 applies to yield hereditary.)

4.18B COROLLARY

Any right bounded pre-Dedekind Cohen ring is Noetherian and hereditary.

Proof. Apply 4.18A.

4.18C COROLLARY

Let R be a prime FPF Cohen ring. Then, R is right Noetherian and right hereditary iff right pre-Dedekind.

Proof. The necessity follows from the fact that in a prime right Noetherian ring, every right ideal ≠ 0 is faithful, and finitely generated, hence a generator of mod-R by FPF. The converse is obtained from 3.22 (which implies that every ideal ≠ 0 is a finitely generated projective right R-module) and 4.18.

4.18D COROLLARY

Let R be a prime FPF Cohen ring. Then R is right Noetherian and right hereditary iff every ideal of R is a finitely generated right ideal.

Proof. For then every ideal ≠ 0 is a finitely generated faithful right ideal, hence generates mod-R, so 4.18C applies. The necessity is trivial.

4.19A LEMMA

A right Goldie right bounded prime ring R with RRM is right Noetherian.

Proof. It suffices to prove that any essential right ideal I is finitely generated inasmuch as any right ideal is a direct summand of an essential right ideal. Since R is right Goldie, then 1.12A implies that I contains a finitely generated essential right ideal I', and then I' contains an ideal $A \neq 0$. Since R is prime, then A is an essential right ideal, hence R/A is right Artinian, hence right Noetherian, and this implies the a.c.c. on right ideals of R contained between I and I'. Thus, since I' is finitely generated, then so is I proving the lemma.

4.19B COROLLARY

A prime FPF ring R with RRM is right Noetherian and right hereditary.

Proof. By 3.16B, R is Goldie, and by 1.3B, R is right bounded, hence the lemma implies that R is right Noetherian. Now any prime RRM ring is Cohen, and since every right ideal is finitely generated, then 4.18D implies the desired conclusion.

4.20 THEOREM

A prime ring R is a bounded Dedekind ring iff R is an FPF ring with left and right restricted minimum conditions.

Proof. Sufficiency follows from 4.6 and the theorem of Chatters and Weber; and the necessity from 4.6 and 4.19B.

The proof of the next theorem is based on an axiomatization of Lenagan's proof, as well as employing Lenagan's theorem 4.2B. Fully right bounded means that every factor ring is right bounded, and fully bounded means both right and left.

4.21 THEOREM

If R is a prime Goldie ring in which ideals $\neq 0$ are invertible, then the following are equivalent:

4.21.1 R is a right bounded Dedekind ring.

4.21.2 R is fully right bounded.

4.21.3 R <u>is a left bounded Dedekind ring</u>.

4.21.4 R <u>is a fully left bounded ring</u>.

<u>Proof</u>. First assume that R is fully right bounded. By 4.2C, any prime ideal $P \neq 0$ is maximal. Since R/P is then simple, and right bounded by hypothesis, then every essential right ideal = R/P, that is R/P is the only essential right ideal, and hence R/P is simple Artinian; R is thereby a Cohen ring. Since every ideal $\neq 0$ is invertible, every ideal is finitely generated projective in mod-R by 3.23.2, so 4.18 implies that R is (hereditary and) Noetherian on the right. Then Lenagan's theorem 4.2B applies to yield the conclusion: R is Noetherian and hereditary on both sides, hence R is Dedekind and bounded on both sides by 4.2B.

The rest follows from 4.10 and a bit of symmetry inasmuch as any right bounded Dedekind prime ring R is CFPF, hence any factor ring is FPF, hence bounded by 4.6, so R is fully bounded.

4.22 COROLLARY

<u>If</u> R <u>is a Goldie prime ring in which ideals</u> $\neq 0$ <u>are invertible, then</u> R <u>is Dedekind iff</u> R/P <u>is Goldie for any prime ideal</u> P.

<u>Proof</u>. The proof is by "proof theory", namely looking at what Lenagan really <u>uses</u> for the proof of his theorem (4.2B). If c is any element of R which maps onto a regular element $\bar{c}$ of $\bar{R} = R/P$, where P is any prime ideal $\neq 0$, then $\bar{c}\bar{R} = \bar{R}$ a fact which Lenagan proves without recourse to the assumption (made in 4.2B) that R is right Noetherian. This assumption is used by Lenagan for the first time in order to obtain that $\bar{R}$ is Goldie, in which case we know that $\bar{R} = Q(\bar{R})$ inasmuch as every regular element $\bar{c}$ is a unit. Thus, $\bar{R} = Q(\bar{R})$ is a semisimple Artinian ring, proving that R is a Cohen ring. Then, the rest follows as in the proof of 4.21.

4.22A COROLLARY

For a bounded prime Goldie ring R, *the f.a.e.:*

1. R *is pre-Dedekind and Noetherian*
2. R *is pre-Dedekind and* R/P *is Goldie for all prime ideals* $P \neq 0$
3. R *is Dedekind.*

Proof. By the proof of 4.2C, every ideal $\neq 0$ of R is invertible when R is pre-Dedekind (as in 1 and 2), so 4.22 applies to give the equivalence of 1.2, and 5.

The next corollary is related to propositions 2.1C and 3.25, and in fact, once we prove R is FPF is a corollary of them, and the fact (to-be proven) that R is Noetherian.

4.22B COROLLARY

A semiperfect pre-Dedekind bounded ring R *is Dedekind, and isomorphic to a full* $n \times n$ *matrix ring* D_n, *over a local right and left principal ideal domain* D.

Proof. By 4.14A, R is actually Prüfer. Actually, the proof of 4.14A shows that every ideal $\neq 0$ is invertible. In any case, R is FPF by 4.12, so by 2.1C, $R = D_n$, where D is a right and left valuation domain. Then, every ideal J' of R has the form J_n for an ideal J of D, so the ideals $\neq 0$ in D are also invertible, i.e. finitely generated. By 4.2C, every prime ideal P of D is maximal. But this implies that $P = \text{rad } D$ is the only prime ideal. Since D is local, this implies that D is Cohen, so that 4.22 applies: D is Dedekind (and hence, so is R). In particular, D is Noetherian, so every onesided ideal is principal.

The next theorem generalizes the last corollary to the case when R is semiprime.

4.22C THEOREM

A ring R *is a finite product of right bounded Dedekind prime rings iff* R *is a fully right bounded semiprime ring in which essential ideals are invertible.*

Proof. If $R = \Pi_{i=1}^{n} R_i$ ia a finite product of

DPR's, and if I is any ideal which is an essential right ideal, then $I_i = I \cap R_i$ is a nonzero ideal of R_i, hence invertible in $Q(R_i)$, with inverse I_i', and then $I = I_1 + \ldots + I_n$ is invertible in $Q(R) = Q(R_1) \oplus \ldots \oplus Q(R_n)$, with inverse $I^{-1} = I_1' + \ldots + I_n'$. Moreover, since R_i is right bounded by hypothesis, then R_i is fully right bounded by 4.10 which implies that R_i is right CFPF. Since any right FPF ring is right bounded by 1.3B, this gives us that R_i is fully right bounded. Thus, if J is any essential right ideal of $R/I \approx R/I_1 + \ldots + R/I_n$, then $J \cap R/I_i$ is an essential in R/I_i, hence contains a nonzero ideal B_i, and then J contains a nonzero (in fact essential) ideal $B = B_1 + \ldots + B_n$. Thus, R is fully right bounded, as required.

Conversely, as in the proof of 4.21, R has the a.c.c. on essential ideals, and every essential ideal is the finite product of maximal ideals. Thus, every essential prime ideal P is maximal, and since a simple right bounded ring must be semisimple, we have that R/P is Artinian. Since every essential ideal is projective (since invertible), then every right ideal containing an essential ideal is projective by a theorem of Zaks [71, Lemma 1, p. 443] (Also see Faith [75, Corollary 6]).

<u>4.23</u> <u>THEOREM</u>

<u>If</u> R <u>is a prime Goldie right FPF ring, then every ring</u> A <u>between</u> R <u>and</u> $Q(R)$ <u>is right FPF</u>.

<u>Proof</u>. As $Q(R)$ is an essential estension of A on both sides, A is a nonsingular ring with maximal quotient ring $Q = Q(R)$, so A is Goldie and prime along with R. Moreover, by 3.18, any finitely generated right ideal $\neq 0$ generates mod-A, so A is right FPF by 4.7.

This theorem generalizes the classical results on overrings of Prüfer and Dedekind domains.

The next theorem is similar (and the proof is similar) to Theorem 3.24.

4.24 PROPOSITION

If R is a (right) Goldie semiprime ring in which (right) faithful, or equivalently (left or right) essential, ideals are invertible (resp. generate mod-R), then R is a finite product of prime rings in which nonzero ideals are invertible (resp. right generators). In case R is Noetherian (Goldie), and ideals are (right) bounded, then R is a finite product of Dedekind (resp. right FPF) prime rings, and hence R is then (right) FPF. Moreover, the complement ideals, the idempotent ideals, and the ring direct factors of R all coincide.

Proof. For an ideal P the conditions right or left faithful or essential are equivalent. To see this suppose P is right faithful and $P \cap K = 0$ some right (left) ideal K. Then $KP = 0$ $(PK = 0)$ so $(KP)^2 = 0$ $((PK)^2 = 0)$, so since R is semiprime $K = 0$. Conversely, if P is essential, and $PK = 0$ for some right ideal K, $(P \cap K)^2 = 0$ implies $K = 0$ since P is essential and R is semiprime. We prove the last statement first.

Suppose that R is right Goldie semiprime, and let B be any ideal $\neq 0$. The right annihilator C of B is also the left annihilator since R is semiprime, and hence C contains any right complement K of B. However, since semiprimeness of R forces $B \cap C = 0$, then C is the right complement of B, and hence $A = B \oplus C$ is an essential right ideal, hence faithful, and therefore (by the hypothesis) generates mod-R. Let A^* denote the R-dual of A contained in the quotient ring Q (as in the proof of 3.21). Then $A^*A = R$, hence

(1) $$R = A^*B + A^*C$$

We shall prove:

(2) $$R = A^*B + C .$$

Now $D = A^*C \cap B$ is "square-zero", so R semiprime implies $D = 0$. But $A^*C \supset C$, and C is the right complement of B, so that $A*C = C$, and therefore (2) holds. Since R is semiprime, this is a ring product. Also, since any ideal which is a right complement is the right complement of an ideal, hence, as we have shown, is a direct summand of R. Furthermore, $B = B^2$ implies via (2) that $B = A*B^2 = A*B$, and then (2) shows that B is a direct factor of R. The converse is trivial: any direct factor is an idempotent ideal, and a right complement.

Since R is right Goldie, then R is a finite product of directly indecomposable rings, each of which inherits the hypothesis on R. Hence assume that R itself is directly indecomposable. Then (2) shows that $R = A*B$ for any nonzero ideal, and hence any nonzero ideal B is a generator. Therefore, if R is Goldie (on both sides) and right bounded, then R is right FPF by Theorem 4.7, that is, every directly indecomposable factor of R is right FPF, and hence then so is R.

In case the assumptions are placed on both sides, then every ideal $\neq 0$ in R would be a generator on both sides, and hence invertible by 3.21.3, so then boundedness an Noetherian implies R is Dedekind by 4.2. This completes the proof.

5 SELF-INJECTIVE FPF RINGS, THIN RINGS, AND FPF GROUP RINGS

In this section we prove the splitting theorem for FPF rings. They split into essentially singular rings and nonsingular rings. The theorem that semi-perfect Noetherian FPF rings are finite products of Dedekind prime rings and Quasi Frobenius rings illustrates the splitting theorem quite well. We will give two proofs of this fact here and give a complete description of Noetherian CFPF rings. The FPF condition is not left right symmetric, i.e. left FPF does not imply right FPF. (It is not even known if non-singular right FPF implies left FPF). It is not known if one sided FPF and Noetherian is enough to achieve a splitting into a product of Dedekind domains and Quasi Frobenius rings. We then examine rings of finite width, i.e. rings for which D(R) is finite (see remarks following Theorem 3.7B). Such rings will be called thin rings. We show, theorem 5.11, that if a ring is self-injective and right FPF, then it is thin and that the converse holds for bounded self-injective rings.

5.1 SPLITTING THEOREM (Faith [79b], Page [83b])

If R is right FPF, then there exists a two sided ideal A which is a direct summand of R as a right ideal such that $Z_r(R)$ the right singular ideal, is essential in A and R/A is right FPF and nonsingular. Moreover, if R is also left FPF, then $R = {}^{\perp}A \oplus A$ as rings.

In order to prove the theorem we need a pair of lemmas.

5.1.A LEMMA

Let R be any ring. Let A be an ideal (two-sided) in R which is closed as a right ideal i.e. $Z_r(R/A) = 0$. If $A + {}^{\perp}A = R$ then, $({}^{\perp}A)^{\perp} = A$.

Proof. We know $({}^{\perp}A)^{\perp} \supset A$ so suppose B is a right ideal such that $B \cap A = 0$. We have $1 = a + y$ with $a \varepsilon A$ and $y \varepsilon {}^{\perp}A$. Since $B \subset ({}^{\perp}A)^{\perp}$, ${}^{\perp}AB = 0$, so $ab = b \varepsilon A \cap B = 0$. Hence $B = 0$. Since A is closed, $({}^{\perp}A)^{\perp} = A$.

5.1B LEMMA

Let R be right FPF. Let $A = \{x \varepsilon R | \ x + Z_r(R) \ \varepsilon \ Z_r(R/Z_r(R))\}$. Let D be a right ideal such that $D \cap A = 0$. Then, D contains no non-trivial square zero right ideals.

Proof. It is well known and not too difficult to see that A is closed. Suppose $B \subset D$ and $B^2 = 0$. Let C be a right ideal maximal w.r.t. $C \cap B^{\perp} = 0$. If H is a two sided ideal and $H \subset C$, $BH = 0$, since $BH \subset B \subset B^{\perp}$ and $BH \subset C$. This says $H \subset B^{\perp}$ so $H = 0$. But then R/C is faithful, hence a generator. In particular R/C must generate C. Now $B^{\perp}$ embeds as an essential submodule of R/C under the natural map and because $B^{\perp}$ is a two-sided ideal, if we take a map $f: R/C \to C$ we see that the image of $B^{\perp}$ in R/C is contained in the kernel of f, i.e. the image of f in C is singular. This implies C is singular. That means $C \subset A$. But $DA = 0$ so in particular $BC = 0$. It must be that $C = 0$. But then $B^{\perp}$ is essential in R, so $B \subset Z_r(R) \subset A$ and hence $B = 0$.

Now to the proof of the theorem. Let A be the closure of the right singular ideal as in lemma 5.1B. We first claim $A + A^{\perp} = R$. To see this take B to be a right ideal maximal w.r.t. $A \cap B = 0$. Then, because R is right FPF, $R = {}^{\perp}A + {}^{\perp}BR$ because $R/A \oplus R/B$ is faithful and ${}^{\perp}A + {}^{\perp}BR$ is the trace of $R/A \oplus R/B$ in R. Suppose $xB = 0$. Then $x(A + B) \subset A$. Because of the choice of B, A + B is an essential right ideal so $(x + A) \varepsilon Z_r(R/A) = 0$ by the choice of A. So $x \varepsilon A$. But then ${}^{\perp}B \subset A$, so $BR \subset A$ and so

$R = A + {}^{\perp}A$. If we write $1 = a + a_1$ where $a \varepsilon A$ and $a_1 \varepsilon {}^{\perp}A$, then $a^2 = a$ and $A = aR$, so A is a direct summand as a right ideal. This is the first part of the theorem.

If we take $R = aR \oplus (1 - a)R$ and use Lemma 5.1B we see that R/A is semiprime. If we can show R/A is FPF, then R/A is nonsingular by Theorem 3.3. Let M be a finitely generated faithful right R/A module. Form $A \oplus M$ as R-modules (give M the natural R structure). Now $M^{\perp} = A$ so $(A \oplus M)^{\perp} = A \cap A^{\perp}$. Using Lemma 5.1A we see that $0 = R^{\perp} = (A + {}^{\perp}A)^{\perp} = A^{\perp} \cap ({}^{\perp}A)^{\perp} = A^{\perp} \cap A$. So $A \oplus M$ is finitely generated and faithful over R. But trace $A \subset A$ since $Z_r(R)$ is essential in A; so M must generate R/A as an R-module, because $Z_r(R/A) = 0$ as an R-module (as an R/A module, too). It follows that M generates R/A as an R/A module, also. So R/A is right FPF.

In order to prove the last statement we have seen that $A^{\perp} \cap A = 0$. But this implies that $A^{\perp} + (A^{\perp})^{\perp} = R$ when R is left FPF. We claim $(A^{\perp})^{\perp} = A$. Now if $B = {}^{\perp}A \cap (A^{\perp})^{\perp}$, $B \cap A = 0$ and $B^2 = 0$ so by Lemma 5.1B, $B = 0$. So $A^{\perp} \cap (A^{\perp})^{\perp} = R$ as two sided ideals. It follows that $A^{\perp}$ is generated by a central idempotent and we can take $\bar{R} = R/A = (A^{\perp})^{\perp}$. But in $\bar{R}$, $\bar{A}$, the image of A, is faithful. Also $\bar{A} = \bar{a}\bar{R}$ as we saw in the previous part of the proof, so $\bar{A}$ is a generator of $\bar{R}$. But $\bar{A}$ can only generate $\bar{A}$, so $\bar{A} = \bar{R}$, i.e. $A = (A^{\perp})^{\perp}$ as desired.

5.2 EXAMPLE (Page [83b])

We will give an example of a right FPF ring which is not left FPF and which shows Theorem 5.1 is best possible. Let Q be a non-semisimple injective regular commutative ring and M an essential maximal right ideal of Q. Let $S = Q/M$ and $D = \mathrm{End}_Q(S)$. Let R be the ring of lower triangular matrices of the form $\begin{pmatrix} q & 0 \\ s & d \end{pmatrix}$ with $q \varepsilon Q$, $s \varepsilon S$ and $d \varepsilon D$. The right singular ideal is the matrices of the form $\begin{pmatrix} 0 & 0 \\ s & 0 \end{pmatrix}$ and the left singular ideal is zero. Also, the ring is right self-injective but not left self-injective. The ring R is also right strongly bounded and hence right

FPF (cf. see Corollary 5.11D).

To see that R is not left FPF notice that the closure of the right singular ideal is e_2R, and e_2R is not a ring direct summand of R. By the theorem, R can not be left FPF. This answers question 5 of Faith [79b].

Faith conjectured there that if R is right FPF, then the maximal right quotient ring is self-injective. (See Open Problems appended at the end.) It is known that this only happens if the maximal quotient ring is the injective hull of R, see ARMC. Since the hull of R is R, this cannot be left self-injective (cf. Corollary 5.11D). It was also conjectured in Faith [79b] that the right maximal quotient ring is the classical right quotient ring. This remains unanswered. For commutative rings Faith [82a] has shown the quotient ring is both injective and a flat epimorph. Notice in 5.2 that R is in a sense as close to commutative as one can get. It's "off by the one element $\begin{pmatrix} 0 & 0 \\ 1 & 0 \end{pmatrix}$". Along the lines of splitting theorems Faith [79b] gives the following application:

5.3 ENDO'S THEOREM

Any commutative Noetherian FPF ring is a product $R_1 \times R_2$ where R_1, is a product of Dedekind rings and R_2 is Quasi-Frobenius.

If we drop commutativity but add semiperfect we obtain:

5.4 THEOREM (Faith [76c])

A ring R is a Noetherian semiperfect FPF ring iff R is a finite product of bounded Dedekind semiperfect rings and Quasi-Frobenius rings.

Proof. By 5.1 we can decompose an FPF ring into a product $R_1 \times R_2$ with R_1 semiprime and R_2 with essential left or right nil ideal. By 4.17 R_1 is Dedekind. By Corollaries 2.13 and 2.21 and Theorem 2 of Chatters and Robson [80] R_2 is Artinian. But Artinian FPF rings are Quasi-Frobenius.

We will give an alternative proof using the following theorem of Robson [74], and theorem 5.6 below which is of interest in its own right. (See ART, Chapter 20)

5.5 THEOREM (Robson [74])

Let R be a Noetherian ring with maximum nilpotent ideal N. Then R is a finite product of Noetherian semiprime rings and Artinian rings iff $cN = N = Nc$ for every $c \in R$ which is regular modulo N.

The proof also requires the next theorem.

A right annulet is principal if it has the form $x^{\perp}$ for some $x \in R$. A radical ideal is one contained in rad R.

5.6 THEOREM (Faith [76c])

If R is a right FPF ring with a.c.c. on principal right annulets, then for every radical ideal A of R, if $c \in R$ maps onto a regular element $\bar{c}$ of $\bar{R} = R/A$, then $\exists\, n > 0$ such that $c^{n\perp}$ contains no nonzero ideal of R, and $c^n R$, hence cR generates mod-R. Furthermore, if R is strongly right bounded, e.g. if R is semiperfect and selfbasic, then $c^{\perp} = 0$.

Proof. Choose n such that $d = c^n$ has the same right annihilator as d^2. We may suppose $n = 1$, and prove that $c^{\perp}$ does not contain an ideal $\neq 0$. Now $c^{\perp} \cap cR = 0$, so by 3.1A setting $Q = (c^{\perp})R$, we have that $R = Q + {}^{\perp}cR$. However, left regularity of $\bar{c}$ in $\bar{R}$ implies that ${}^{\perp}cR \subset A$, and hence that $R = Q + A$. However, if $q \in Q$ and $a \in A$ are such that $1 = q + a$, then by the fact that $A \subset \text{rad } R$, we have that $q = 1 - a$ is a unit, hence $Q = R$. This proves, as in the proof of 3.1A that the trace ideal of $R/c^{\perp}$ is R, that is $cR \approx R/c^{\perp}$ generates mod-R, and hence that $R/c^{\perp}$ is a faithful right module. Thus, $c^{\perp}$ can contain no nonzero ideal. Note: in the general case, $c^n R$ generates mod-R, but $cR \to c^n R \to 0$ is exact, so cR does, too. For the last sentence, apply 2.1A.

Now for the proof of 5.4:

Proof. By 3.4.2 a finite product of rings is FPF (resp. right bounded) iff each of the canonical factors is. Moreover, semiperfect rings are also closed under finite products and direct factors. Therefore, since bounded Dedekind prime rings are (C)FPF by 4.10, and QF rings are (not only FPF but) PF by 1.8, then the sufficiency is proved.

Conversely, since any Noetherian semiprime FPF ring is a finite product of bounded Dedekind rings by 4.10 and any Artinian FPF ring is QF by 1.9 (applying 1.6 and 1.8), then the necessity follows once the hypotheses of Robson's theorem are verified, namely: $cN = N$, for all c in R which are regular modulo N. (Then $Nc = N$ follows by symmetry.) To prove this we first assume that R is selfbasic, and apply 5.6 to conclude that c is a right regular element of R. Therefore the right ideal $I = cr + N$ is faithful, and hence generates mod-R, and hence $I \approx R$ by 2.1B. Therefore, we may write $cR + N = dR$, for some right regular element d. Then $N \subset dR$, and $\bar{d}\bar{R} = \bar{c}\bar{R}$, so, since $\bar{R}$ is semiprime Goldie, d is regular modulo N which implies that $a \in N$ whenever $da \in N$, and hence that $N = dN$. Then, we can write $d = cx + dy$, for some $x \in R$ and $y \in N$ (in as much as $dR = cR + dN$). Hence we have $d(1-y) = cx$, and since $1 - y$ is a unit, we have $d \in cR$, proving that $cR = dR$, and hence $cN = dN = N$. Furthermore, by symmetry, $N = Nc$, as required. This proves that the basic ring of R has the desired structure, but then so does R.

<u>5.7</u> <u>THEOREM</u> (Faith [76c])

<u>Any Noetherian CFPF ring</u> R <u>is a finite product of bounded Dedekind prime rings and primary-decomposable Artinian serial rings</u>. <u>Conversely</u>.

<u>Proof</u>. The converse follows from the fact that any finite product of CFPF rings is CFPF (Theorem 3.4.2).

Let N be the maximal nilpotent ideal. We shall apply Robson's theorem as in the proof of 5.4. Therefore, let $c \in R$ be such that $\bar{c}$ is regular in $\bar{R} = R/N$. By CFPF, $\bar{R}$ is FPF, hence by 4.10, $\bar{R} = \bar{R}_1 \times \dots \times \bar{R}_n$, where $\bar{R}_i$ is a bounded Dedekind prime ring, $i = 1, \dots, n$. Since $\bar{c}\bar{R} \cap \bar{R}_i$ is an essential right ideal of $\bar{R}_i$, then by the fact that $\bar{R}_i$

is bounded, we have that $\bar{c}\bar{R} \cap \bar{R}_i$ contains an ideal $\bar{B}_i \neq 0$, $i = 1,\ldots,n$. Since $\bar{R}_i$ is prime, the $\bar{B}_i$ is essential in $\bar{R}_i$, and hence $\bar{c}\bar{R}$ contains an ideal $\bar{B} = \bar{B}_1 + \ldots + \bar{B}_n$ which is essential in $\bar{R}$. Therefore, by the Goldie theorems, $\bar{B}$ contains a regular element $\bar{d}$ of $\bar{R}$, and, in fact, we can assume that $d \varepsilon B$, where B is the ideal of R containing N mapping canonically onto $\bar{B}$. By Theorem 5.6, dR generates mod-R, hence B is faithful. Since R is CFPF, this is true for any factor ring R/A, for any ideal $A \subset N$, that is, B/A is a faithful R/A module. Consequently, B/BN is a faithful R/BN module, and this implies that $BN = N$. Therefore, since $cR + N \supset N$, we have that

$$N \supset (cR + N)N = cN + N^2 \supset BN = N$$

that is, $cN + N^2 = N$. If k is the index of nilpotency of N, then $cN^{k-1} = N^{k-1}$.

Then, since $cN^{k-2} + N^{k-1} = N^{k-2}$ we have that

$$cN^{k-2} + cN^{k-1} = N^{k-2} \subset cN^{k-2}$$

hence $cN^{k-2} = N^{k-2}$. An evident induction yields $cN = N$, and by symmetry, $Nc = N$, so Robson's theorem implies that R is a product $R_1 \times R_2$, where R is semiprime and R_2 is Artinian. Then, the argument of Theorem 5.4 yields that R_1 is a finite product of bounded Dedekind prime rings, and R_2 is QF. However, since R_2 is CFPF, then R is CPF by 1.9 (since Artinian) hence R_2 is uniserial by Theorem 1.10.

The converse is obtainable as in the proof of Theorem 5.4.

Theorem 5.7 may be thought of as a generalization or extension of Warfield's Theorem 6.10 of [75] in which he proved for a Noetherian algebra A finitely generated as an R-module over a Noetherian commutative ring R that the f.a.e.:

(i) If I is an ideal such that A/I is Artinian, then A/I is a right and left PIR.

(ii) every f.g. module is a direct sum of a projective module with no simple submodules $\neq 0$ and a finite number of Artinian "homogeneously" serial modules. (This means all the simple factors of the serial modules are isomorphic.)

(iii) A is a product of an Artinian ring which is left and right PIR, and a finite number of maximal orders (in simple algebras) over Dedekind domains.

(iv) every f.g. module over A is balanced, that is, A maps epically onto the Biendomorphism ring of any f.g. module.

A ring A with property (iv) for all right modules is said to be right balanced. (See ARMC, page 250 for several references to these rings and Warfield's theorem.) We say that a ring A is finitely right balanced if (iv) holds for all f.g. right modules. By Morita's theorem 1.10, every right CFPF ring is finitely right balanced, and Warfield's theorem provides a converse for finitely right and left balanced Noetherian algebras A, as stated, inasmuch as (iii) implies that A is CFPF. In fact, Artinian PIR's are CPF by 1.8A and the theorem of Asano stated sup. 1.6, and maximal orders over Dedekind domains are HNP's, hence Dedekind prime rings by 4.1, hence CFPF by 4.10. It would be interesting to determine the Noetherian finitely balanced rings. Conjecture: they are CFPF rings.

THIN RINGS

One of the relevant concepts in the study of FPF rings is the idea of width (see Theorem 3.8A and the definition preceding). Namely:

5.8 DEFINITION

1. Let M be an R-module and let D(M) be called the width of M.
2. If $D(M) < \infty$ we will say M is thin otherwise we will say M is thick.

5.9 PROPOSITION

If M is an R-module, $D(M) = D(E(M))$ where $E(M)$ is the injective hull of M, and Goldie dim M = Goldie dim $E(M)$.

Proof. Let N be any submodule of M and suppose N^n embeds in M. Then surely N^n embeds in $E(M)$. It follows that $D(M) \leq D(E(M))$. Suppose N is a submodule of $E(M)$. Suppose N^n embeds in $E(M)$. Let $\text{Im } N^n = N_1 \oplus N_2 \oplus \ldots \oplus N_n$. Now $N_1 \cap M \neq 0$, since M is essential, and N_2 contains a copy, W_2, of $N_1 \cap M = W_1$. Also, $W_2 \cap M \neq 0$, so let $W_2' = W_2 \cap M$. Clearly W_1 contains a copy of W_2', W_1' say. Now N_3 contains a copy of W_1', W_3 say. $W_3 \cap M = W_3' \neq 0$ and both W_1' and W_2' contain copies of W_3'. So we have W_1'' and W_2'' each isomorphic to W_3'. Continuing in this way we obtain, after n steps, submodules $A_1, A_2, \ldots A_n$ all contained in M, with $A_i \cong A_j$ for all i and j, and $A_i \subset N_i$ for $i = 1, \ldots n$. Thus $D(M) \geq n$. It follows that $D(M) = D(E(M))$.

5.10A PROPOSITION

Let M be an injective R-module with $B \neq \text{End } M_R$, and $J = \text{rad } B_R$ Then:

(1) *M is thin iff B/J is right thin.*

(2) *M has finite Goldie dimension iff B/J is semisimple.*

Proof. Let $M = F^n \oplus X$ for a submodule F of M, where M is a slim injective. Let $e_1, \ldots, e_n$ be idempotents of $B = \text{End}(M_R)$ where $e_i M = F_i$ and $F = F_1 \oplus \ldots \oplus F_n$, with $F_i \cong F$. Taking s_{ij} to be the element of B which interchanges the ith "coordinate" with the jth "coordinate" and leaves everything else fixed we have that $s_{ij} e_i s_{ji} = e_j$, and this gives a B isomorphism of $e_i B$ and $e_j B$ for $i = 1, \ldots, n$. Thus $D(M) \leq D(B)$. On the other hand it is well known that $B/J(B)$ is a right self-injective ring, [ART, 19.27, p 76] and idempotents lift. It follows that if $\bar{e}_i, \ldots, \bar{e}_n$ is a system of orthogonal idempotents of $B/J(B)$, there exists a system $e_1, e_2, \ldots, e_n$ of orthogonal idempotents in B. Since $B/B(J)$ is self-injective $D(B/J(B)) = \sup\{n \mid B \supset \oplus \Sigma_{i=1}^n e_i B,\ e_i^2 = e_i,\ e_i B \cong e_j B,\ e_i e_j = 0\ i \neq j,\ i = 1, \ldots n\}$. The result will follow if we show $e_iM \approx e_jM$ as R-modules. The fact that $e_iB \cong e_jB$ implies there are elements a and b of B so that $e_iB = ae_jB$, and $be_iB = e_iB$. The map e_jbe_i embeds e_iM into e_jM. We have then that e_iM isomorphic to a submodule of e_jM and by symmetry e_jM is isomorphic to a submodule of e_iM. By Bumby [65] we have $e_iM \approx e_jM$.

5.10B COROLLARY

Let M as an R-module and $B = \mathrm{End}(E(M)_R)$. Then $D(M) = D(E(M))$ $D(B/J(B))$.

5.10C COROLLARY

Let M be an R-module. Then M is thin iff M^n is thin for all finite integers n, and $D(M^n) = nD(M)$.

Proof. Since $E(M^n) = (E(M))^n$, $\mathrm{End}(E(M)^n)$ is the $n \times n$ matrix ring over $\mathrm{End}(E(M)) = B$, and the Jacobson radical of a matrix ring is the matrices with entries from the Jacobson radical, the corollary follows.

5.10D COROLLARY

Let R be a right self-injective ring. If S is Morita equivalent to R, then S is thin as a right S module iff R is thin as a right R-module.

Proof. S is after all the endomorphism ring of a finitely generated injective right R-module and is right self-injective, too! So $D(S) = D(S/J(S))$.

We now give a characterization of self-injective right FPF rings.

5.11A THEOREM

Let R be a self-injective (both sides) ring. Then R is right FPF iff R is right thin and Morita equivalent to a right strongly bounded ring S with $D(S) = 1$.

Proof. First assume R is right FPF and self-injective. We know $R/J(R) = \bar{R}$ is a von Neumann regular self-injective ring, that $J(R) = Z(R)$ and $\bar{R}$ is Dedekind finite, by Utumi [65]. We wish to show $\bar{R}$ is FPF. If $\bar{R}$ is not right FPF, then $\bar{R}$ is not of bounded index. Hence

$\bar{R}$ contains idempotents $\{\bar{e}_i\}_{i=1}^{\infty}$, such that $\bar{e}_i R\text{-dim}(R) = i$, $e_i e_j = 0$, $i \neq j \ \forall\ i$ and $j = 1,2,\ldots$ Let $\bar{e} = \Pi \bar{e}_i$ which exists in R since R is right self-injective. We have that $E(\bar{R}\bar{e}\bar{R}) = \bar{A}$ is a two-sided direct summand of $\bar{R}$. So $\bar{R} = \bar{A} \oplus \bar{B}$. Choose right ideals A and B of R so that A maps to $\bar{A}$ and B maps to $\bar{B}$, and lift all the idempotents to R. Now $\bar{A}$ embeds in a product of copies of $\bar{e}\bar{R}$ because $\bar{R}\bar{e}\bar{R}$ is essential in $\bar{A}$. We claim A embeds in copies of eR. To see this let $a \varepsilon A$. Now $E(aR) = fR$ where $f^2 = f$, and $\bar{f} \varepsilon \bar{A}$. We can find a map of $h: \bar{f}\bar{R} \to \bar{e}\bar{R}$ which is not zero. By the injectivity and regularity of $\bar{R}$, this map h is monic on a submodule of $\bar{f}\bar{R}$ which is a direct summand of $\bar{f}\bar{R}$. So $\bar{f}\bar{R} = \bar{f}_1\bar{R} \oplus \bar{f}_2\bar{R}$, $f_1^2 = f_1$, $f_2^2 = f_2$ and h restricted to $\bar{f}_1\bar{R}$ is an isomorphism of $\bar{f}_1\bar{R}$ and a submodule of $\bar{e}\bar{R}$. Since $Z(R) = J(R)$ a standard argument implies h lifts to an isomorphism of f_1R and a submodule of eR. Because aR is essential in fR, $aR \cap f_1R \neq 0$. This gives a map h_1: $aR \to eR$ such that $h_1(a) \neq 0$ and since a is arbitrary, we have the required embedding. Clearly $(eR)^{\perp} \supset A^{\perp}$, so from the embedding $(eR)^{\perp} = A^{\perp}$. Finally, B is generated by an idempotent and so $eR \oplus B$ is faithful and hence a generator. But then $\bar{e}\bar{R}$ is a generator of $\bar{A}$, which is impossible since the ring is Dedekind finite. Since $\bar{R}$ is now right FPF, $\bar{R}$ is of bounded index and we can take $\bar{e} \varepsilon \bar{R}$, where $\bar{e}^2 = \bar{e}$ and lift $\bar{e}$ to e, an idempotent in R, such that $D(\bar{e}\bar{R}) = 1 = D(eR)$. Since eR is projective $S = \text{End}(eR)$ is Morita equivalent to R and it is easy to see that $D(S) = 1$. To see that any self-injective ring R with $D(R) = 1$ and right FPF is strongly right bounded, let $z \varepsilon R$. Then zR is essential in eR with $e^2 = e$. If zR contains no two sided ideals, $eR/zR \oplus (1-e)R$ is faithful and hence a generator. But trace(eR/zR) is contained in $Z(R)$ and, since $D(R) = 1$, the trace of $(1-e)R$ in eR is contained in $Z(eR)$, contradicting the fact that $eR/zR \oplus (1-e)R$ is a generator.

To establish the converse since both FPF and

self-injectivity are Morita invariant properties, we can assume R is strongly bounded and $D(R) = 1$ and R is right self-injective. If M is a finitely generated module over R which is faithful, let $M = \Sigma_{i=1}^{n} m_i R$. Now $\cap_{i=1}^{n} (m_i R)^{\perp} = 0$. Since R is right self-injective we will show M finitely co-generates R. To this end we may as well assume $\cap_{j \in N} (m_j R)^{\perp} \neq 0$ where N is any proper subset of n. Since R is strongly bounded each $m_j^{\perp}$ contains a non-zero two sided ideal B_j and B_j is essential in $m_j^{\perp}$. But $B_j = (m_j R)^{\perp}$, so $\cap_{j=1}^{n} B_j = 0$. Because the B_j's are essential in the $m_j^{\perp}$'s it follows that $\cap_{j=1}^{n} m_j^{\perp} = 0$. This says the map $r \mapsto \Sigma_{i=1}^{n} m_i r \in M^n$ is an embedding.

5.11B COROLLARY

If R is Dedekind finite and right self-injective, then R is right FPF iff R is thin and Morita equivalent to a right strongly bounded ring S with $D(S) = 1$. Moreover, $R/J = \bar{R}$ is FPF (both sides) regular and injective on both sides and $\bar{R}$ is Morita equivalent to $\bar{S} = S/J(S)$.

Proof. The only place we used left injectivity was in showing that R is Dedekind finite.

Therefore, the structure theorem for $\bar{R}$ stated in chapter 3 (theorem 3.14A ff.) applies, hence $\bar{R}$ has bounded index, is left self-injective and is Morita equivalent to a strongly regular ring $\bar{S}$.

Some questions arise from the above: (1) Are there any self-injective rings with $D(R) = 1$ which are not right strongly bounded? (2) Are there any self-injective right FPF rings which are not left FPF?

5.11C COROLLARY

If R is right self-injective and CFPF, then R is thin.

Proof. We now have $R/J(R)$ is regular and FPF hence self-injective by 3.14A hence Dedekind finite and hence that R is Dedekind finite.

5.11D COROLLARY

If R is any right strongly bounded right self-injective ring, then R is right FPF.

In order to show how rings with $D(R) = 1$ play a role like that of the basic rings of Chapter 2 we have the following: (Compare 2.1A)

5.12 PROPOSITION

Let R be a self-injective right FPF ring. Then R contains a subring R_o with R_o Morita equivalent to R and any f.g. faithful R_o-module contains a copy of R_o.

Proof. As in 5.11 we can take $R_o = eRe$ where $D(eR) = 1$. To finish the proof we need to show any f.g. faithful R_o module contains a copy of R_o. We have that $0 \to R_o \to M^{(n)}$ exact for some n since R_o is right FPF. Choose n as small as possible. Let A_i be the kernel of the projection of R_o onto the ith coordinate of $M^{(n)}$. Then $\cap_{i=1}^n A_i = 0$. Let $B = \cap_{i>1} A_i$. Since n is as small as possible $A_1 \cap B = 0$. Now $R_o = e_1R_o \oplus e_2R_o$ where $A_1 \subset e_1R$ and $B \subset e_2R_o$, $e_i^2 = e_i$ $i = 1,2$. We have that $e_2R_o \hookrightarrow M^{n-1}$ and $e_1R_o \hookrightarrow M$. Now think of M as a submodule of M^{n-1}. So we have $e_1R_o \hookrightarrow M^{n-1}$. But $\mathrm{Im}\, e_1R_o \cap \mathrm{Im}\, e_2R_o = 0$ since $D(R_o) = 1$. So there is a sequence $0 \to R_o \to M^{n-1}$ exact. By the minimality of n we must have $n = 1$ and the proposition is proved.

5.13 PROPOSITION

Let R be a self-injective ring (both sides) with $D(R) = 1$. If R is CFPF (both sides) and every homomorphic image is self-injective, then $D(R/A) = 1$ for all two sided ideals A.

Proof. Suppose A is a two sided ideal and $A \subset J$, the Jacobson radical of R, then $D(R/A) = 1$ since idempotents lift. If $A \not\subset J$ we have by 5.1 that $R/A \cap J = \bar{R} = R_1 \times R_2$ with R_1 semi-prime and $J(R_2)$ essential in R_2. Now under the natural map A maps to R_1 for if $a \in A$ and $a \notin J$, $a^2y - a \in J$ for some y and hence $a^2y - a \in A \cap J$, i.e. $(A/A \cap J)\ \mathrm{rad}(R/J \cap A) = 0$. Now since idempotents lift $D(R_1) = 1 = D(R_2)$. We also have $\mathrm{Hom}(R_i, R_j) = 0$, $i \neq j$. We have R_1 strongly regular and $R_1/(A/A \cap J)$ is therefore strongly regular. Finally $R/A = R_1/(A/A \cap J) \times R_2$ and the proposition is proved.

SPLIT-NULL EXTENSIONS

We next show how to construct some self-injective rings and in particular PF rings. Let B be a ring and E a bimodule over B. The semidirect product* of B and E is the set product $B \times E$ made into a ring by $(b_1, e_1)(b_2, e_2) = (b_1 b_2, b_1 e_2 + e_1 b_2)$ and addition component wise. We begin with the main lemma used in the constructions.

5.14 LEMMA

Let R be a ring, let E be an ideal which is its own left annihilator, ${}^{\perp}E = \{a \in R \mid aE = 0\} = E$, let $B = R/E$. Then E is canonically a B-bimodule. If (14.1) E is injective as a (canonical) right B-module, and (14.2) $B \approx \mathrm{End}E_B$ canonically, then R is right self-injective (= injective mod-R).

Conversely, if R is right self-injective, then for any ideal A, the left annihilator ${}^{\perp}A$ is an injective right R/A-module, and $\mathrm{End}\ {}^{\perp}A_{R/A} \approx R/{}^{\perp\perp}A$ canonically. Thus, in this case, any ideal E satisfying $E = {}^{\perp}E$ satisfies (14.1) and (14.2).

Proof. Let F be the injective hull of R in mod-R, and let

$$F_1 = \mathrm{ann}_F E = \{x \in F \mid xa = 0\ \forall a \in E\}.$$

Then, F_1 is a right B-module, and $E = {}^{\perp}E$ is an injective right B-module by (1.1). Since every B-submodule of F_1 is an R-submodule, then F_1 is an essential extension of $F_1 \cap R = E$ as an R-module, hence as a B-module, so injectivity of E in mod-B implies that $F_1 = \mathrm{ann}_F E = E$. Thus, if $y \in F$, then $yE \subset \mathrm{ann}_F E = E$, so y induces an endomorphism $b \in B' = \mathrm{End}\ E_R = \mathrm{End}\ E_B$. Now every $r \in R$ induces an endomorphism $r_s \in \mathrm{End}\ E_B$ via left multiplication;

---

*Also called the split-null or trivial extension of B by E.

hence $B = R/{}^{\perp}E = R/E$ embeds in B' canonically. Since $B \approx B'$ canonically by the assumption (14.2) there exists $r \varepsilon R$ such that

$$yx = b(x) = r_s x = rx \quad \forall x \varepsilon E$$

so

$$(y - r)x = 0 \quad \forall x \varepsilon E;$$

hence

$$y - r = c \varepsilon \operatorname{ann}_F E = E \subseteq R.$$

Therefore, $y = r + c \varepsilon R$, $\forall y \varepsilon F$, proving that $F = R$ is injective. In this case, for *any* ideal A, ${}^{\perp}A$ is an injective right R/A-module (e.g. ART, p.66, Prop. 12) and every $b \varepsilon \operatorname{End} A_R$ is induced by an element $r \varepsilon R$; hence $R/{}^{\perp}A \approx \operatorname{End} A_R$. Also, $R/{}^{\perp\perp}A \approx \operatorname{End} {}^{\perp}A_R = \operatorname{End} {}^{\perp}A_{R/A}$, canonically. Taking $A = E = {}^{\perp}E$, we have the stated properties (1) and (2).

5.15 THEOREM (Faith [77])

Let $R = (B,E)$ *be the semidirect product of a bimodule* E *over a ring* R. *Thus,* $a(xb) = (ax)b$ *for all* $a,b \varepsilon B$ *and* $x \varepsilon E$, *and in* $R = B \times E$ *addition is componentwise, and multiplication is defined by*:

(15.1) $$(a,x)(b,y) = (ab, ay + xb).$$

(*The ring* R *is* $\approx$ *the ring of all* 2×2 *matrices* $\begin{pmatrix} a & x \\ 0 & a \end{pmatrix}$, *with* $a \varepsilon B$, $x \varepsilon E$, *under ordinary matrix operations*.) *Then*:

(15.2) R *is right self-injective iff* E *is injective in mod-B, and* $B = \operatorname{End} E_B$ *canonically*.

(15.3) R *is a right injective cogenerator in mod-R* (= R *is right PF*) *iff* E *is an injective cogenerator of mod-B satisfying* $B = \operatorname{End} E_B$ *canonically*.*

---

*Commutative FPF rings $R = (B,E)$ are characterized in Faith [83/84].

(15.4) <u>Assuming</u> (15.3), <u>then</u> R <u>is left PF iff</u> E <u>is an injective cogenerator of</u> B-mod, <u>and</u> $B = \text{End}_B E$ <u>canonically</u>.

<u>Proof</u>. (15.2). Identify E with $E_1 = \{(0,x) \mid x \in E\}$ in R, and B with $B_1 = \{(b,0) \mid b \in B\}$. Clearly, $B \approx B_1 \approx R/E_1$ (under $b \mapsto (b,0)$) and ${}^{\perp}E_1$ in R is E_1 if E is a faithful left B-module. Thus, assuming E_B injective and $B = \text{End}E_B$, that is, assuming (14.1) and (14.2), we have R is injective by Lemma 5.14. The converse also comes from Lemma 5.14.

(15.3). Assume that R is right PF (= pseudo-Frobenius). As remarked <u>sup</u>. 1.7A, and injective right R-module E is cogenerating iff every simple right R-module embeds in E. Since R is a right injective cogenerator ring by assumption, every simple right R-module $V \hookrightarrow R$. Now, since $J = \text{rad } R$ contains any square-zero (or nilpotent or nil) ideal, then $J \supset E_1$; hence $R/J \approx B/\text{rad } B$, and every simple right R-module $V = R/M$ corresponds to a simple right B-module $V' = B/M'$. Since V embeds in R, then V' embeds in R. If $v \in R$ and $v = (b,x) \neq 0$ generates V, then $b = 0 \Rightarrow V \supset E$, and $b \neq 0 \Rightarrow \exists 0 \neq (0,y) \in E$ such that $0 \neq (b,x)(0,y) \in V \cap E$; hence $V \cap E = V \subset E$ in both cases. This proves that every simple B-module V' embeds in E. Since E is injective by (15.2), this proves that E is cogenerating in mod-B. Moreover, $B = \text{End } E_B$ via (15.2).

These remarks also suffice for the converse of (15.3), since E cogenerating means every simple B-module V' embeds in E; hence every simple R-module V embeds in E. Thus, if E is an injective cogenerator in mod-B, and $B = \text{End } E_B$, then R is injective by (15.2), hence cogenerating inasmuch as every simple right R-module V embeds in $E_1 = (0,E) \subset R$.

<u>Proof</u> of (15.4). Let R be left PF. Since E is an injective cogenerator of mod-B (by the assumption (15.3)), then E is faithful as a right B-module (see, e.g. ART, p.92 II4(a)); hence $E^{\perp} = E_1$ follows, so E_1 is an injective left B-module, where $B = R/E_1$, and it is easy to

see that $E \approx E_1$ is actually an injective cogenerator of B-mod: If V is a simple left B-module, then V is a simple left R-module, so $V \subset R$. But $E_1 V = 0$, since V is a B-module, so $V \subset E_1^{\perp} = E_1$ making E_1 a cogenerator of B-mod. (Cf. ART, p. 199, Exercise 1.) Conversely, if E is an injective cogenerator of B-mod, and $B = \mathrm{End}_B E$, then by the right-left symmetry of Lemma 5.14, R is left self-injective, hence cogenerating inasmuch as every simple left B-module V embeds in $E_1 = (0,E) \subset R$.

5.15A COROLLARY.

Let $R = (B,E)$ be the semidirect product of a ring B and B-bimodule E. Then: R is cogenerating(both sides) iff $\mathrm{End}(E_B) = B = \mathrm{End}({}_B E)$ and E is a cogenerator over B on both sides.

Proof. A ring R is cogenerating on both sides iff R is PF on both sides (1.7C). Therefore, Theorem 5.15 applies.

The corollary shows: E a strongly balanced cogenerator over B does not imply that $R = (B,E)$ is cogenerating. However, it can be shown that R is then FP-injective, and, moreover, every ideal of R is an annihilator ideal.

Every known example of a right PF ring is left PF.

5.15B COROLLARY.

If every right PF ring is left PF, then a bimodule E over a ring B satisfies (14.1) and (14.2) iff it satisfies the left-right symmetry (14.1) and (14.2). (Compare Theorem 1.9).

Proof. This follows from Theorem 5.15 part 15.4.

If E is injective in mod-B and $B = \mathrm{End}\, E_B$ canonically, is then E injective in B-mod, that is, as a left B-module? And if so, is $B \approx \mathrm{End}\, {}_B E$ the endomorphism ring of the left B-module E?

The next corollary shows that in (15.3) for a

commutative ring B, we may restrict ourselves to a local ring B, and the next theorem shows that similarly in (15.2) for Noetherian commutative B, we may restrict ourselves to the case where B is local, and furthermore, in order that $R = (B,E)$ be injective, it is necessary that E be not only injective but a cogenerator.

5.16A COROLLARY

If E is a B-bimodule satisfying (15.3), then B is semiperfect, and E is a finite direct sum of indecomposable injectives. Therefore, there are only finitely many nonisomorphic simple B-modules, and E has finite socle.

Proof. Since $R = (B,E)$ is right PF, then R is semiperfect (1.7A) and the rest follows from this.

5.16B THEOREM (Faith [77])

Let B be a commutative Noetherian ring with an injective module E, such that $B = \text{End}_B E$. Then $B = \Pi_{i=1}^n B_i$ is a finite product of complete local rings, and $E = \oplus \Sigma_{i=1}^n E_i$, is the smallest injective cogenerator of B_i, $i = 1,\ldots,n$. Thus, E is the smallest injective cogenerator of B.

Proof. Since B is Noetherian, E is a finite coproduct $E = \Pi_{i=1}^n E_i$ of indecomposable injectives. Since each E_i has a local endomorphism ring, the finite Krull-Schmidt theorem holds, and so B is a semilocal ring. Moreover, since idempotents lift modulo the radical (ART, p. 45, 18.26), then $B = \Pi_{i=1}^n B_i$, where $B_i = e_i B e_i \approx \text{End}_B E_i$ is a local ring, and $e_i^2 = e_i \varepsilon B$ is the projection idempotent, $i = 1,\ldots,n$. Hence, we may assume E is indecomposable and B local. By Matlis' Theorem [58] in order that B be complete it is necessary and sufficient to show that E is the injective hull of $V = B/\text{rad } B$. By the Matlis-Utumi Theorem, $J = \text{rad } B$ is the set of all b such that $bx = 0$ for some $x \neq 0$. Since J is f.g., and E is uniform, then

$W = \mathrm{ann}_E J \neq 0$. Thus, W is an R/J-module, hence is semisimple (= a direct sum of simples), whence simple by uniformity, so $W \approx R/J \hookrightarrow E$. Then, E is the injective hull of $V = R/J$, as required.

5.17A COROLLARY

If $B = \mathrm{End}_B E$ is a commutative local ring with f.g. radical J, and E injective, then $E = E(B/J)$ is the injective hull of B/J. So E is a cogenerator in mod-B.

Proof. Same.

5.17B COROLLARY

If the semidirect product ring $R = (B,E)$ of a Noetherian commutative ring B and module E is self-injective, then R is an injective cogenerator, and a finite product of local injective cogenerators.

Proof. By Theorem 5.15 $B = \mathrm{End}_B E$ canonically, and E is an injective module, so Theorem 5.16B applies, and the rest is easy.

An application of Theorem 5.15 and Matlis' Theorem [58] yields:

5.17C THEOREM (Faith [77])

If B is a Noetherian ring, and $E = E(B/\mathrm{rad}\ B)$ the injective hull, then $R = (B,E)$ is injective iff B is complete. (Then R is PF.)

5.18A PROPOSITION

A semidirect product ring $R = (B,E)$ is a right VR iff B is a right VR, E is uniserial, and $bE = E$ $\forall b \neq 0$, and $b \in B$.

Proof. If R is a right VR, then $B \approx R/(0,E)$ is a right VR, and $E \approx (0,E)$ is uniserial. If $0 \neq b \in B$, then $(b,0)R \not\subset (0,E)$; hence

$$(b,0)R = (bB,0) + (0,bE) \supset (0,E),$$

so bE = E. The converse follows by reading up.

A VD is a domain which is a VR. For simplicity, for 5.18B - 5.20 we shall assume that B whence R is commutative.

5.18B COROLLARY.

Let E be a faithful B-module. Then R = (B,E) is a VR iff B is a VD and E is a uniserial divisible B-module.

Proof: Immediate.

5.18C. COROLLARY.

Let E be a torsion free module over a domain B. Then R = (B,E) is a VR iff B is a VD and E is a uniserial injective B-module. In this case R is injective iff B = End(E).

Proof. Any torsion free divisible module over a domain is injective, so apply the corollary. (conversely, any injective module is divisible.) The last sentence follows from Theorem 5.15.

5.19A THEOREM (Faith [77])

Let R = (B,E) be a semidirect product ring. The f.a.e.::

1) R is a PFVR.

2) B is an almost maximal valuation domain (AMVD), E = E(B/rad B) is the injective hull of B/rad B, and B = $\text{End}_B E$.

3) B is a local domain such that E = E(B/rad B) is uniserial and strongly balanced.

4) B is an MVD and E = E(B/rad B) is strongly balanced.

Proof. By Gill's theorem [71] a local ring B is AMVR iff E(B/J) is uniserial, where J = rad B. Thus, using Theorems 5.15 and 5.18A (2)<=>(3) follows. Moreover, (1)<=>(3) by 5.18C and Corollary 5.17A and (2)<=>(4) by a theorem of Vamos [75].

<u>5.19B</u> <u>COROLLARY</u>.

<u>If</u> B <u>is a Noetherian local domain, and</u> $E = E(B/J)$, <u>then the semi-direct product ring</u> $R = (B,E)$ <u>is an injective VR iff</u> B <u>is a complete discrete valuation domain</u>. <u>In this case</u> R <u>is PF</u>.

<u>Proof</u>. Follows from 5.19A and Matlis' theorem [58] (since B is a Noetherian VD).

<u>5.20</u> <u>EXAMPLE</u> (Levy [66]).

<u>A Noncogerating Injective Local Ring</u>.

Let A denote the ring of all formal power series $\sum c_a x^a$, where $c_a \in \mathbb{R}$ and $a \in \mathbb{R}$, with the usual addition and multiplication. As stated the proper ideals of A are in 2-1 correspondence with $\mathbb{R}$, namely, for each $b \in \mathbb{R}$, we have the "closed" (= principal) ideal

$$(x^b) = \{x^b u \mid u \in A\}$$

and the "open" ideal

$$(x^{>b}) = \{x^c u \mid u \in A,\ c > b\}.$$

(In particular, rad $A = (x^{>0})$). Levy [66] proved that every proper factor ring $R = A/I$, $I \neq A, 0$, is self-injective. Clearly no proper factor ring $A/I, I \neq 0$, contains a minimal ideal, hence A/I is injective but not PF.

If R is any QF ring, and G a finite group, then the group ring RG is QF by a theorem of Connell [63]. The extension of this result to right PF rings is Theorem 5.22. The general result for FPF rings is false even for $R = Z$ essentially because $\mathbb{Z}G$ is not hereditary and every Noetherian semiprime FPF ring is not only hereditary, but also a maximal order (Theroem 4.11; Cf. 4.1).

A right ideal I of R is <u>dense</u>, written $I <_r R$, provided that R is a rational extension of I in mod-R. As stated following Corollary 1.12A' this happens iff $\mathrm{Hom}_R(A/I,R) = 0$ for all right ideals $A \supset I$. Thus, if R has no dense right ideals, then, in particular, every simple right R-module embeds in R. Since every nonzero module M has a nonzero map into any cogenerator, then a right cogenerator ring, in particular a right PF ring R has no dense right ideals. Conversely, if R is injective in mod-R, then R is a cogenerator that is, right PF, if R has no dense right ideals. This will be used to show that right PF is a property of R inherited by any group ring RG of a finite group G over R. Louden's theorem below is needed. The following corollary first appeared in Faith [76].

5.21 THEOREM (Louden [76])

Let R be a subring of a ring S such that S is a Frobenius extension of R (= there is an (R,S) - bimodule isomorphism $\mathrm{Hom}_R(S,R) \approx S$). If S is finitely generated as a module over R by elements $\{x_i\}_{i=1}^n$ such that $x_iR = Rx_i \forall i$, e.g. if $S = RG$, with $|G| < \infty$, any dense right ideal D of S contracts to a dense right ideal in R. Moreover, $\mathrm{Hom}_R(S, \ )$ mod-R $\hookrightarrow$ mod-S preserves injective hulls.

Proof. When a ring S has the stated generation over a subring R, then, $\mathrm{Hom}_R(S, \ )$ preserves essential extensions, equivalently, injective hulls. To see this suppose A is an essential submodule of the right R-module B. Let $0 \neq f \varepsilon \mathrm{Hom}_R(S,B)$. For x_1, if $f(x_1) = 0$ let $y_1 = 1$. If $f(x_1) \neq 0$ choose $r_1 \varepsilon R$ so that $0 \neq f(x_1)r_1 \varepsilon A$. Now $x_1r_1 = y_1x_1$ for some y_1. If $f(y_1x_2) = 0$ let $y_2 = 1$. If $f(y_1x_2) \neq 0$ choose $r_2 \varepsilon R$ so that $0 \neq f(y_1x_2r_2) \varepsilon A$ and y_2 so that $x_2r_2 = y_2x_2$. Note that $f(y_1y_2x_1) = f(y_1x_1w)$ for some $w\varepsilon R$, so that $f(y_1y_2x_1) \varepsilon A$. Inductively choose $y_3, y_4 \cdots, y_n$ as above and let $z = y_1y_2 \cdots y_n$. We have $f_z \neq 0$ and $f_z(x_i) = f(zx_i) \varepsilon A$ for $i = 1, \cdots, n$. It follows that $f_z(s) \varepsilon A$ for all $s \varepsilon S$ and that $\mathrm{Hom}_R(S,A)$ is essential in $\mathrm{Hom}_R(S,B)$. Next, if E denotes the injective hull of A in mod-R, then $S' = \mathrm{Hom}_R(S,E)$ is an essential extension of $T = \mathrm{Hom}_R(S,A)$, as we have shown. Moreover, by injectivity of E and the natural isomorphisms

$$\mathrm{Hom}_S(X,S') \approx \mathrm{Hom}_R(X \otimes_S S,E) \approx \mathrm{Hom}_R(X,E)$$

we see that S' is the injective hull of T in mod-S. Thus if $E = E(R)$ denotes the injective hull of R in mod-R, then $S' = \mathrm{Hom}_R(S,E)$ must be the injective hull of $S \approx \mathrm{Hom}_R(S,R)$ in mod-S, hence by the adjointness of $\mathrm{Hom}_R(S,\ )$ and the tensor (over S) by S,

$$D_r < S \Rightarrow 0 = \mathrm{Hom}_S(S/D,S') \approx \mathrm{Hom}_R(S/D,E)$$

Since $R/(D \cap R) \to S/D$ in mod-R, then injectivity of E implies that $\mathrm{Hom}_R(R/(R \cap D),E) = 0$, that is, $R \cap D_r < R$.

5.22 COROLLARY

Under the same hypotheses, then R right PF implies that S is right PF.

Proof. Since R is injective in mod-R, and $\mathrm{Hom}_R(S,\ )$ preserves injective hulls from mod-R to mod-S, then $S \approx \mathrm{Hom}_R(S,R)$ in mod-S is injective. Furthermore, S has no dense right ideals $\neq S$, since R has none, so S is right PF.

For commutative FPF rings we do have:

5.23A THEOREM (Faith [82a])

If R is commutative, injective, and if G is a finite group then RG is injective and FPF (both sides).

Proof. Let M be a f.g. faithful RG-module. Then $M = x_1R + \ldots + x_nR$ is a f.g. R-module for suitable $x_1,\ldots,x_n$ in M. Let $S = \mathrm{End}(M_{RG})$. Then R embeds canonically in S so M is f.g. by $x_1,\ldots,x_n$ as a left S module, so $S^n \to M \to 0$ is exact in S-mod. Apply ${}_S(\ ,M)$ to obtain $0 \to {}_S(M,M) \to {}_S(S^n,M) = M^n$ exact in mod - A were $A = \mathrm{End}({}_SM)$. So A embeds in M^n. But RG embeds in A canonically and RG is injective by Connell's theorem [63]. So we have a split sequence $0 \to RG \to M^n$ and thus RG is FPF.

5.23B COROLLARY

If R is commutative FPF, then the classical ring of quotients of RG for G a finite group is FPF and hence is the maximal ring of quotients.

Proof. This follows from Faith [82a] theorem 5.1, which implies that $Q = Q_{c\ell}(R)$ is self-injective and FPF, hence by the theorem so is QG. Obviously every regular element of RG is regular, hence a unit in QG; consequently $Q_{c\ell}(RG) \subset QG$, so injectivity of QG, and the fact that RG is an order in QG imply that $Q_{c\ell}(RG) = QG$.

In view of 5.23A and B, the following application of the splitting theorem is noteworthy.

5.24 THEOREM

If R is a self-injective right and left FPF ring, then $R \approx R_1 \times R_2$ where R_1 is a self-injective von Neumann regular ring of bounded index and R_2 is a self-injective ring with essential Jacobson radical, with $D(R_2) < \infty$.

Proof. By Utumi [65] the Jacobson radical and singular ideal are the same for a self-injective ring. The result then follows by 5.1 and 5.11.

The next result is a simple application of averaging argument widely applied in group rings.

5.25 LEMMA

Let R be a ring and G a finite group with order of G a unit in R. Then for an RG module M, $Z_r(M_{RG}) = Z_r(M_R)$.

Proof. By Lorenz and Passman [80, Lemma 2], the RG-essential right ideals of RG are R-essential and the result follows.

5.26 THEOREM

If R is self-injective right FPF and G a

finite group such that the order of G _is a unit in_ R, _then_ RG _is right FPF_.

Proof. We have that RG is self-injective by Connell [63]. Also, since $D(R) < \infty$ we have $D(RG) < \infty$, too. Now let $e = e^2$ be an idempotent of RG so that $D(eRGe) = 1$ and eRGe is Morita equivalent to RG. We claim eRGe is strongly bounded. To this end let H be a right ideal in eRGe = S. Since S is self-injective there is in S an idempotent $f = f^2$ so that fS is an essential extension of H. Now if H does not contain a non-trivial ideal, $fS/H \oplus (1-f)S = M$ is a faithful finitely generated S module. Since RGeRG = RG, $M \otimes_S eRG = fRG/HRG \oplus (1-f)eRG = \tilde{M}$ is an RG-module and is faithful since the two sided ideals of RG correspond to the two sided ideals of S bijectively. Next note that $\tilde{M}$ is a finitely generated faithful R module, hence must be a generator of mod-R. We have also that fRG/HRG is RG-singular, hence by Lemma 5.24 R-singular. It follows that (e-f)RG is a generator of mod-R since the trace of fRG/HRG in R is contained in the singular ideal which is the Jacobson radical of a self-injective ring. The trace of (e-f)RGe in fRGe is contained in the singular submodule of fRGe since $D(S) = 1$. By the usual Morita equivalence it follows that the trace of (e-f)RG in fRG is also contained in the singular submodule of fRG as an RG-module. However, as R-modules, this can not be the case, i.e. there is a map $\lambda:(e-f)RG \to fR$ such that image λ is not contained in $Z_r(fR)$. (Note $Z_r(fR) \neq 0$ since $fR(RG) = fRG$ is projective and not singular.) Lift λ to $\tilde{\lambda}(x) = \Sigma_{g\varepsilon G} f\lambda(xg)g^{-1}$. As usual $\tilde{\lambda}$ is an RG map of (e-f)RG to fRG. Now, fRG is R isomorphic to order G copies of fR, so since $\tilde{\lambda}(x)$ is singular there is an essential right ideal of R such that $\tilde{\lambda}(x)E = 0$, and this implies $\Sigma_{g\varepsilon G} f\lambda(xEg)g^{-1} = 0$, which implies $f\lambda(xEg) = 0$ for each $g\varepsilon G$. In particular we also have that $\lambda(x)E = 0$, a contradiction. Theorem 5.11A now gives the result.

5.27 EXAMPLE

Note that if R is self-injective and has no order of G torsion then the order of G is a unit in R. If we drop the self-injectivity in the above the result fails. To see this let T be the ring of integers localized at 2, and $R = T[i,\sqrt{3}]$, where $i^2 = -1$. Next take G the group with three elements and form RG. Now R is a valuation ring hence faithful finitely generated ideals are $\approx R$, and so is FPF, by Faith [79b]. A bit of calculation shows there are idempotents in $Q[i,\sqrt{3}]G$ not in RG. The ring RG is non-singular hence by 3.28, RG can not be FPF.

SUMMARY OF RESULTS NONCOMMUTATIVE RINGS

In Chapters 3 and 4 the structure of nonsingular one sided FPF rings is largely given. It is shown that they are nonsingular iff they are semiprime and that they are nonsingular on both sides. The maximal quotient ring is shown to be a two sided maximal quotient ring and FPF on both sides. The embedding of the ring in its maximal quotient ring is a flat epimorphism. Then von Neumann regular FPF rings are shown to be precisely the self-injective (both sides) rings of bounded index, and hence FPF on both sides. If besides nonsingular, the condition of finite Goldie dimension is imposed, then an FPF ring must be a semiprime Goldie ring on both sides. If the further restriction of A.C.C. on left and right ideals is added, then the ring is a bounded Dedekind domain and CFPF and conversely.

We do not know of a nonsingular FPF ring which is not semihereditary. If all nonsingular FPF rings are semihereditary, they are Baer rings as are all the finite matrix rings of said rings. The converse also holds, namely, if a ring is an FPF Baer ring, as well as the finite matrix rings over it, the ring is semihereditary. For the case of commutative FPF rings we do know if the nonsingular ones are hereditary (see the next section of this summary). A right Noetherian nonsingular FPF Cohen ring is hereditary. Infact, any right bounded nonsingular Cohen ring with the maximal right quotient ring the left maximal quotient ring for which ideals $\neq 0$ are generators is Noetherian and hereditary. A prime right FPF ring with restricted right minimum condition is also hereditary.

Finally we have that for prime right FPF rings, every ring between the ring and its maximal quotient ring is right FPF.

The results for singular FPF rings are not so definitive. For a semiperfect right FPF ring we have that the ring is a direct sum of uniform principle indecomposable right ideals and, moreover, that the basic module is isomorphic to a direct summand of any finitely generated faithful module, and the basic ring is strongly right bounded. If in addition the ring modulo the Jacobson radical is prime, the ring is the matrix ring over local right FPF rings, and in case the ring is prime, these local rings are right and left valuation domains. A right FPF semiperfect ring with nil radical is self-injective and for a local right FPF ring it is self-injective iff the radical consists of zero divisors. A left perfect right FPF ring is right PF. Finally, a semiperfect right CFPF is a finite product of full matrix rings over right duo right valuation right σ-cyclic right CFPF rings.

In the more general setting, all two sided FPF rings are the product of a semiprime ring and a ring with essential singular ideal (either singular ideal). If the latter ring is Noetherian, it is quasi Frobenius, and of course conversely. This also leads to a description of the Noetherian CFPF rings as finite products of bounded Dedekind prime rings and primary-decomposable Artinian serial rings.

Self-injective thin rings have a "basic ring" which is Morita equivalent to the ring much as do semiperfect rings. A self-injective ring is right FPF iff this basic ring is right strongly bounded. As in the case for semiperfect rings, the basic ring is a direct summand of any finitely generated faithful module. A self-injective FPF ring is a product of a von Neumann regular self-injective ring and a self-injective ring with essential singular ideal. If a ring is commutative self-injective and FPF and we take a finite group, then the group ring is FPF but for general FPF rings this is false even if the order of the group is a unit in the ring. However, if the ring is self-njective and the order of the group is a unit in the ring, then the group ring is FPF.

COMMUTATIVE FPF RINGS.

Since the results for commutative FPF rings are more decisive than for noncommutative rings we include here a brief summary of these results (which can be found in Faith [79B, 81]) to provide additional insights and motivation.

In what follows all rings are commutative. Q will be the maximal ring of quotients of a ring R and Q_c the classical ring of quotients of R. J will denote the Jacobson radical of Q_c.

Theorem: *A commutative ring* R *is FPF iff* Q_c *is injective and faithful finitely generated ideals are projective.*

The proof depends on a "converse" of a theorem of Azumaya which states that f.g. faithful projective modules are generators over commutative rings. The 'converse" is that any submodule of Q which is a generator is f.g. projective.

For reduced rings:

Theorem: *A reduced commutative ring* R *is FPF iff* R *is a semihereditary ring with* Q_c *injective.*

Theorem: *If* R *is FPF and* $P = J \cap R$, *then* $R = R/P$ *is reduced, hence semihereditary with* $Q_c(R) = Q_c/J$.

Partial converse: *If* Q_c *is injective, and* R *is FPF then* $R' = R + J$ *is FPF.*

A commutative ring R is *quotient injective* if Q_c is injective. A ring is *fractionally self-injective* (FSI) if every factor ring is quotient injective. Vamos [77] characterized the FSI rings as finite products of *almost maximal rings* $R_1, \ldots, R_n$ where each R_i is either a valuation ring, an h-local domain or a torch ring, $i = 1, \ldots, n$. A domain is called *h-local* if each ideal is contained in only finitely many maximal ideals and each prime ideal is contained in only one maximal ideal. A ring R is almost maximal if each R_P is an AMVR for all maximal ideals P. R is a torch ring if (i) R has at least two maximal ideals, (ii) R has a unique minimal prime $P \neq 0$ and P is uniserial and for an ideal I either $I \subset P$ or $I \supset P$ (P is a waist), (iii) R/P is an h-local domain.

Theorem: _A commutative ring_ R _is CFPF iff_ R _is FSI_.

The FSI rings arise in the classification of FGC = (σ-cyclic) rings see Brandal [79], Vamos [77] or R. and S. Wiegand [77].

A _Bezout_ ring is a ring for which all f.g. ideals are principal. Then R is FGC iff R is an FSI Bezout ring.

Corollary. R _is FGC iff_ R _is a CFPF Bezout ring_.

Corollary. _The following conditions on a local ring_ R _are equivalent_. (i) R _is FGC_; (2) R _is MVR_; (3) R _is FSI_; (4) R _is CFPF_.

Corollary. _A domain_ R _is FPF iff_ R _is Prüfer_; _and_ R _is CFPF iff_ R _is an almost maximal_ h-_local_ _(Prüfer) domain_.

In the study of FPF rings, we encounter what are called sandwich rings: R is a _sandwich ring_ if R contains the radical $J = \text{rad } Q_c$ of Q_c. This is a trivial concept if Q_c is semiprimitive since then $J = 0$, but often, e.g. when Q is local, it is quite meaningful. Thus, a local ring is FPF iff R is a quotient-injective sandwich ring such that R/J is a VR.

Another nice property of FPF rings: they are integrally closed. Thus, any FPF ring R contains every idempotent of Q, and indeed, every nilpotent element. Thus, if Q has nil radical, then R must be a sandwich ring. Actually, this shows that whenever R is a sandwich ring such that R/J is FPF, then R is FPF.

OPEN QUESTIONS

1. a) Are right FPF rings thin? All known right FPF rings R are right thin; e.g., commutative or semiperfect or nonsingular or right self-injective right FPF rings are. For any ring R, the f.a.e.: (RT1) R is right thin; (RT2) $Q^r_{c\ell}(R)$ is right thin; (RT3) $Q^r_{max}(R)$ is right thin; (RT4) R has right thin injective hull E(R); (RT5) $\Lambda = \text{End } E(R)_R$ is right thin; (RT6) $\bar{\Lambda} = \Lambda/\text{rad }\Lambda$ is right thin; (RT7) $\bar{\Lambda}$ is right FPF; (RT) $\bar{\Lambda}$ has bounded index.
   b) Characterize FPF rings. (Commutative FPF rings are characterized in Faith [82a].)

2. Characterize when the trivial extension $R = (B,E)$ of a B-bimodule E is right FPF. (See theorem 5.15 where this is done for PF, and Faith [83/84] where this is done for commutative R and faithful E.

3. Is the center of an FPF ring necessarily FPF?

4. Does right PF ⇒ left PF? Consult Osofsky [66], particularly statement (P) on p. 385 loc. cit.

5. Does right FPF ⇒ left FPF for nonsingular or semiperfect? If so, then, by a theorem of Faith [77], 4 has a yes answer.

6. If R is right FPF is $Q_{c\ell} = Q^r_{max}(R)$?
   Answer is yes if R is right nonsingular (semiprime) by Chapter 3, or if R is commutative by a theorem of Faith [82a]. We conjecture this holds in general, at least for 2-sided FPF rings assuming $Q_{c\ell}(R)$ exists. We also conjecture $Q_{c\ell}(R)$ is self-injective for 2-sided FPF rings.

7. Is every right FPF Kasch ring right PF? (Cf. Faith [79b], esp. Propositions 1F and 1G.) This is equivalent to the requirement that R be right self-injective. Thus, since $R = Q_{c\ell}(R)$ for a commutative Kasch ring, the answer is yes for commutative rings by Faith [82a].

8. Characterize FPF group rings RG, for a necessarily FPF ring R and group G. If R is right FPF, then the finite group ring RG need not be right FPF; e.g. $\mathbb{Z}G$ is never FPF when $\infty > |G| > 1$. Is it, however, when $|G|$ is a unit? (If R is right PF, then so is RG by a theorem of K. Louden (cf. Chapter 5), without assuming that $|G|$ is a unit. Compare 5.23 A and B).

9. If R is left perfect right FPF, then R is right PF by Tachikawa's theorem [69]. Is R then QF? If R is left PF, then the answer is yes by a theorem of Osofsky [66] (for then there is a Morita duality context ${}_R R_R$).

10. If R is right and left perfect right (F)PF, is QF? Cf. #9.

11. Is a nonsingular FPF ring R necessarily semihereditary? The answer is yes in case (1) R is semiperfect (Chapter 3), or (2) R is commutative, (Faith [79b, 82a]), or R is Noetherian. See theorems 3.31.1, 4.10, 4.18c and 4.19.

12. Find an example of a ring R with genus R ≠ Big genus R.

13. If R is a commutative ring, and if every ideal of R has commutative endomorphism ring, then is $Q_{max}(R)$ injective? (The converse holds, as we have noted in Faith [82a]. Also note that for any R, every faithful ideal has commutative endomorphism ring.) This is a problem appended at the end of Faith [82a].

14. If R is right FPF, and if G is a finite group of automorphisms, then is the Galois subring A = RG right FPF? For a commutative ring R, an affirmative answer appears in Faith [82c] when R is finitely generated projective over A. The general questions remain open: does R FPF ⇒ R^G FPF? See the article cited for some more general sufficient conditions. We conjecture the answer is negative in general.

15. Characterize semiperfect (resp. right Noetherian) right FPF rings, or any permutation therof. (In Chapter 5, we determined all semi-perfect, right <u>and</u> left Noetherian right <u>and</u> left FPF rings, without characterizing either semiperfect and/or right Noetherian right FPF rings!)

BIBLIOGRAPHY

References in the text, e.g. Morita [58] refer to a work by Morita listed below published in 1958. If more than one appears in a year, then small letters will distinguish them, as [76a], [76b], in the order they appear.

Akasaki, T., "Idempotent ideals in integral group rings", Archiv. der Math., 24 (1973), 126-8

Asano, K., "Uber verallgemeinerte Abelsche Gruppen mit hyperkomplexen Operatorenring und ihre Anwendungen", Japan. J. Math., 15 (1959), 231-53

__________, "Uber hauptideal ringe mit kettensatz", Osaka Math. J., 1 (1949), 52-61

__________, "Zur Arithmetik in Schiefrungen II", J. Inst. Polytech., Osaka City Univ. Ser-A Math 1, 1-27 (1950)

Azumaya, G., "A duality theory for injective modules", Amer. J. Math., 81 (1959), 249-278

__________, "Completely faithful modules and self-injective rings", Nagoya J. Math., 17 (1966), 249-78

Bass, H., "Finitistic dimension and a homological generalization of semiprimary rings", Trans. Amer. Math. Soc., 95 (1960), 466-88

__________, "On the ubiquity of Gorenstein rings", Math. Zeitschr., 82 (1963), 8-28

__________, Algebraic K-Theory, Benjamin, New York, 1968

Beachy J.A., "Bicommutators of cofaithful, fully divisible modules", Can. J. Math., 23 (1971), 202-13

Beachy J.A., and Blair, W.D., "Rings whose faithful left ideals are cofaithful", Pac. J. Math., 1975

Boyle, A.K., see Goodearl, K.R.

Brandal, W., "Commutative rings whose finitely generated modules decompose", Lecture Notes in Math., Vol. 723, 1979

Brungs, H.H., "Generalized discrete valuation rings", Can. J. Math., 21 (1969) 1404-8

__________, and Torner, G., "Chain rings and prime ideals", preprint, University of Alberta, Edmonton, 1977.

Bumby, R.T., "Modules which are isomorphic to submodules of each other", Arch. der Math., 16 (1965) 184-185

Camillo, V., and Fuller, K.R., "Rings who faithful modules are flat", Archiv. Math., (Basel), 27 (1976) 522-25

Cartan, H., and Eilenberg, S., Homological Algebra, Princeton 1956

Cateforis, V.C., "On regular self-injective rings", Pac. J. Math., Vol. 30, No. 1 (1969), 39-45

________, "Flat regular quotient rings", Trans. Amer. Math. Soc., 138 (1969), 241-49

________, and Sandomierski, F.L., "The singular submodule splits off", J. Alg., 10 (1968), 149-165

Chase, S.U., "Direct products of modules", Trans. Amer. Math. Soc., 97 (1960) 457-73

Chatters, A.W. and Robson J.E. "Decomposition of Orders in semiprimary rings", Comm. in Alg., 8 (6) (1980) 517-532

Cohen, I.S., and Kaplansky, I., "Rings for which every module is a direct sum of cyclic modules", Math. Z., 54 (1951) 97-101

Connell, I., "On the group ring", Canad. Math. J., 15 (1963) 650-85

Eilenberg, S. see Cartan, H.

Eisenbud, D., "Subrings of Artinian and Noetherian Rings", Math. Ann., 185 (1970) 247-349

________, and Robson, J.C., "Hereditary noetherian prime rings", J. Algebra, 16(1970) 67-85

________, and Griffith, P., "Serial rings", J. Algebra, 17 (1971) 389-400

Endo, S., "Completely faithful modules and quasi-Frobenius algebras", J. Math. Soc. Japan, 19 (1967) 437-456.

Faith, C., "Rings with ascending chain condition on annihilators", Nagoya Math. J., 27 (1966) 179-91

________, "On Kothe rings", Math. Ann., 164 (1966) 207-12

________, "Lectures on injective modules and quotient rings", Lecture Notes in Mathematics, 49, Springer-Verlag, Berlin, Heidelberg, and New York, 1967

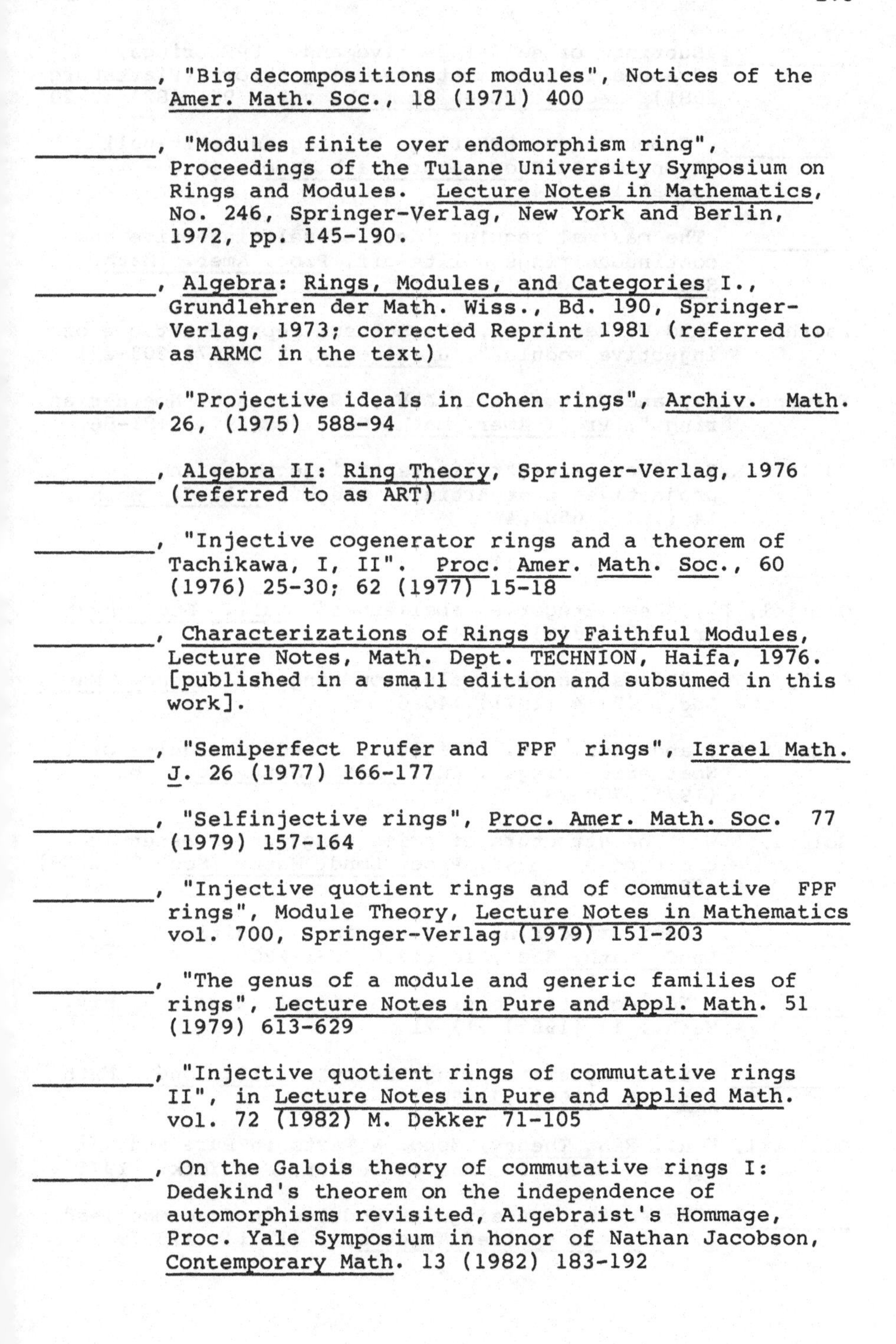

__________, "Big decompositions of modules", Notices of the Amer. Math. Soc., 18 (1971) 400

__________, "Modules finite over endomorphism ring", Proceedings of the Tulane University Symposium on Rings and Modules. Lecture Notes in Mathematics, No. 246, Springer-Verlag, New York and Berlin, 1972, pp. 145-190.

__________, Algebra: Rings, Modules, and Categories I., Grundlehren der Math. Wiss., Bd. 190, Springer-Verlag, 1973; corrected Reprint 1981 (referred to as ARMC in the text)

__________, "Projective ideals in Cohen rings", Archiv. Math. 26, (1975) 588-94

__________, Algebra II: Ring Theory, Springer-Verlag, 1976 (referred to as ART)

__________, "Injective cogenerator rings and a theorem of Tachikawa, I, II". Proc. Amer. Math. Soc., 60 (1976) 25-30; 62 (1977) 15-18

__________, Characterizations of Rings by Faithful Modules, Lecture Notes, Math. Dept. TECHNION, Haifa, 1976. [published in a small edition and subsumed in this work].

__________, "Semiperfect Prufer and FPF rings", Israel Math. J. 26 (1977) 166-177

__________, "Selfinjective rings", Proc. Amer. Math. Soc. 77 (1979) 157-164

__________, "Injective quotient rings and of commutative FPF rings", Module Theory, Lecture Notes in Mathematics vol. 700, Springer-Verlag (1979) 151-203

__________, "The genus of a module and generic families of rings", Lecture Notes in Pure and Appl. Math. 51 (1979) 613-629

__________, "Injective quotient rings of commutative rings II", in Lecture Notes in Pure and Applied Math. vol. 72 (1982) M. Dekker 71-105

__________, On the Galois theory of commutative rings I: Dedekind's theorem on the independence of automorphisms revisited, Algebraist's Hommage, Proc. Yale Symposium in honor of Nathan Jacobson, Contemporary Math. 13 (1982) 183-192

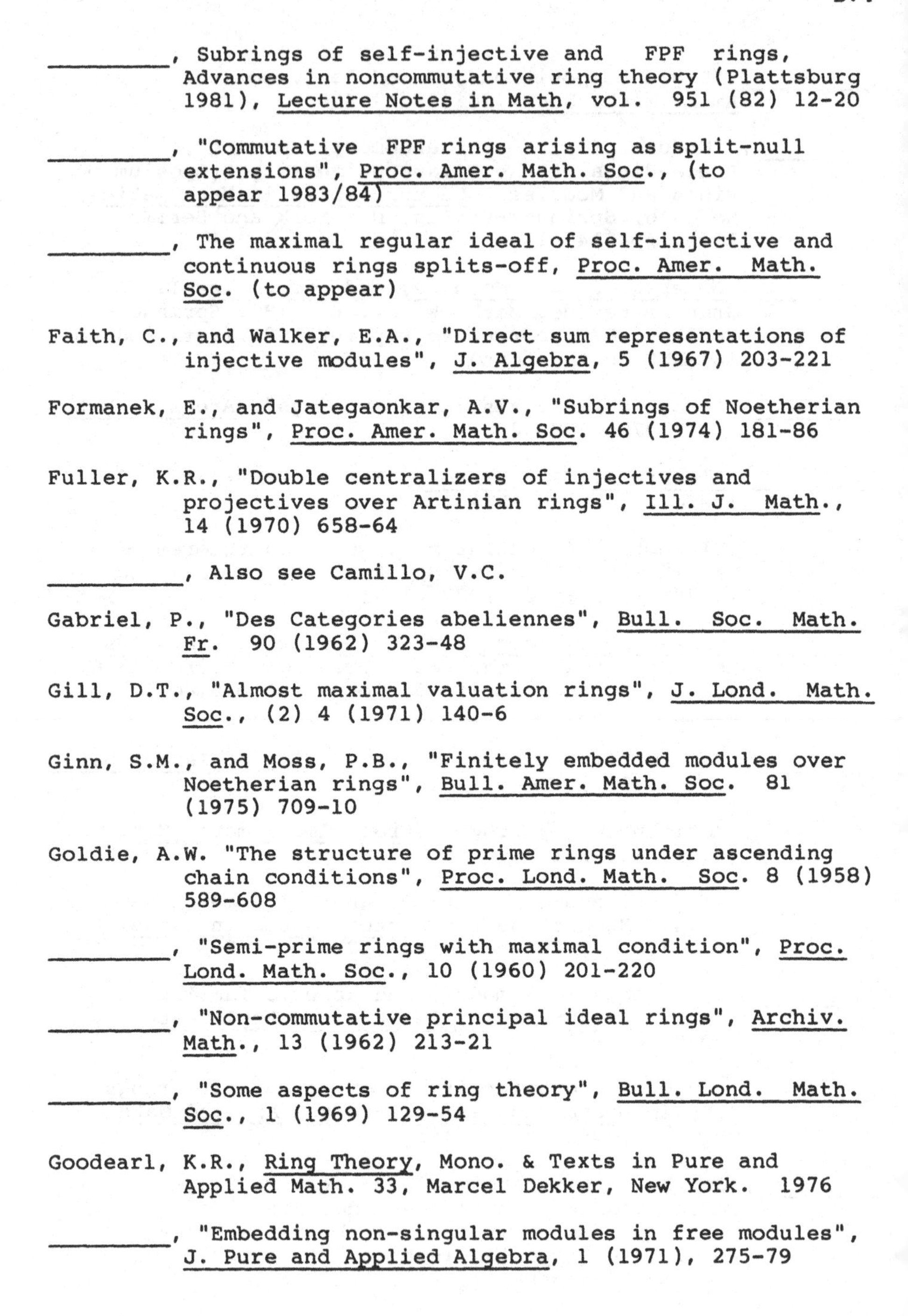

________, Subrings of self-injective and FPF rings, Advances in noncommutative ring theory (Plattsburg 1981), Lecture Notes in Math, vol. 951 (82) 12-20

________, "Commutative FPF rings arising as split-null extensions", Proc. Amer. Math. Soc., (to appear 1983/84)

________, The maximal regular ideal of self-injective and continuous rings splits-off, Proc. Amer. Math. Soc. (to appear)

Faith, C., and Walker, E.A., "Direct sum representations of injective modules", J. Algebra, 5 (1967) 203-221

Formanek, E., and Jategaonkar, A.V., "Subrings of Noetherian rings", Proc. Amer. Math. Soc. 46 (1974) 181-86

Fuller, K.R., "Double centralizers of injectives and projectives over Artinian rings", Ill. J. Math., 14 (1970) 658-64

________, Also see Camillo, V.C.

Gabriel, P., "Des Categories abeliennes", Bull. Soc. Math. Fr. 90 (1962) 323-48

Gill, D.T., "Almost maximal valuation rings", J. Lond. Math. Soc., (2) 4 (1971) 140-6

Ginn, S.M., and Moss, P.B., "Finitely embedded modules over Noetherian rings", Bull. Amer. Math. Soc. 81 (1975) 709-10

Goldie, A.W. "The structure of prime rings under ascending chain conditions", Proc. Lond. Math. Soc. 8 (1958) 589-608

________, "Semi-prime rings with maximal condition", Proc. Lond. Math. Soc., 10 (1960) 201-220

________, "Non-commutative principal ideal rings", Archiv. Math., 13 (1962) 213-21

________, "Some aspects of ring theory", Bull. Lond. Math. Soc., 1 (1969) 129-54

Goodearl, K.R., Ring Theory, Mono. & Texts in Pure and Applied Math. 33, Marcel Dekker, New York. 1976

________, "Embedding non-singular modules in free modules", J. Pure and Applied Algebra, 1 (1971), 275-79

__________, and Boyle, A.K., "Dimension theory for non-singular injective modules", Memoirs, Amer. Math. Soc. #177 (1976)

Griffith, P., "On the decomposition of modules and generalized left uniserial rings", Math. Ann., 184 (1970) 300-308

__________, Also see Eisenbud, D.

Hall, M., Jr., "A type of algebraic closure", Ann. of Math., 40 (1939) 360-9

Handelman, D., and Lawrence, J., "Strongly prime rings". Trans. Amer. Math. Soc., 211 (1975) 209-23

Hosaka, H., and Ishikawa, T., "On Eakin-Nagata's Theorem", J. Math., Kyoto Univ. 13 (1973), 413-416.

Ikeda, M., "A characterization of quasi-Frobenius rings", Osaka Math. J., 4 (1952), 203-10

Ishikawa, T. See Hosaka, H.

Jacobson, N., Theory of Rings, Surveys of the Amer. Math. Soc., (1943), Providence

__________, "Some remarks on one-sided inverses", Proc. Amer. Math. Soc., 1 (1950), 352-355

__________, Structure of Rings, Amer. Math. Soc., Providence, 1964

Jategaonkar, A.V. See Formanek, G.

Johnson, R.E., "The extended centralizer of a ring over a module", Proc. Amer. Math. Soc., 2 (1951), pp. 891-95

Kaplansky, I., "Maximal fields with valuation", Duke Math. J., 9 (1942), 303-21

__________, "Rings of Operators", W.A. Benjamin, New York (1968),

__________, "Elementary divisors and modules", Trans. Amer. Math. Soc., 66 (1949) 464-91

__________, "Modules over Dedekind rings and valuation rings", Trans. Amer. Math. Soc., 72 (1952) 327-40

__________. Also see Cohen, I.S.

Kato, T., "Selfinjective rings", Tohoku Math. J., 19 (1967) 485-95

Klatt, G.B. See Levy, L.

Kothe, G., "Verallgemeinerte Abelsche Gruppen mit hyperkomplexen Operatorenring", Math. Z., 39 (1935) 31-44

Lambek, J., "Rings and Modules", Blaisdell, New York, corrected reprint Chelsea 1966; 1976

Lawrence, J. See Handelman, D.

Lenagan, T.H., "Bounded Asano Orders are hereditary", Bull. Lond. Math. Soc., 3 (1971) 67-9

Lenzing, H., "Halberblicke Endomorphismenringe", Math. Z., 118 (1970) 219-240

Levy, L.S., "Torsion free and divisible modules over non-integral-domains", Canada J. Math., 15 (1963) 132-51

__________, "Commutative rings whose homomorphic images are self-injective", Pac. J. Math., 18 (1966) 149-53

__________, and Klatt, G.B., "Pre-self-injective rings", Trans. Amer. Math. Soc., 122 (1969) 407-419

Lorenz, M. and Passman, D.S., "Two Applications of Maschke's Theorem", Comm. Alg., 8 (19), (1980) 1853-1866

Louden, K. "Maximal Quotient Rings of Ring Extensions" Pac. J. Math., Vol 62, #2 (1976)489-496

Matlis, E.,"Injective modules over Prufer rings", Nagoya Math. J., 15 (1959) 57-69

__________, "Indecomposable modules", Trans. Amer. Math. Soc., 125 (1966) 147-69

Michler, G.O., "Asano Orders", Proc. Lond. Math. Soc., 19 (1969) 421-43

__________, "Structure of semi-perfect hereidtary rings", J. Algebra, 13 (1969) 327-44

Morita, K., "Duality for modules and its applications to the theory of rings with minimum condition", Sci. Reports, Tokyo Kyoiku Daigaku, 6 (1958) 83-142

Moss, P.B. See Gill, D.T.

Muller, B.J., "On Morita duality", Canada J. Math., 21 (1969) 1338-1347

Nakayama, T., "Note on uniserial and generalized uniserial rings", Proc. Imp. Acad. Tokyo, 16 (1940), 285-89

________, "On Frobeniusean algebras I, II", Ann. of Math., 40(1939), 611-33

Osofsky, B.L.,"Rings all of whose finitely generated modules are injective", Pac. J. Math., 14 (1964) 645-50

________, "A generalization of Quasi-Frobenius rings", J. Algebra, 4 (1966) 373-387

Page, S., "Regular FPF Rings", Pac. J. Math., Vol. 79, No. 1, (1978) 169-176

________, "Corrections and Addendum to "Regular FPF Rings" Pac. J. Math., Vol 97, No. 2 (1981) 488-490

________, "Semiprime and Nonsingular FPF Rings", Comm. Alg. Vol 10 (1982) 2253-2259

________, "Semihereditary and Fully idempotent FPF Rings", Comm. in Alg., 11(3) (1983) 227-242

________, "FPF Rings and Some Conjectures of C. Faith", to appear in Canada Bull. Math. 1983/84

________, "Semiperfect FPF Rings", Proc. Amer. Math. Soc. (to appear 83/84)

Passman, D. See Lorenz, M.

Popescu, N. and Spircu, T., "Quelques observations sur les epimorphismes plat (a gauche) d'anneaux", J. Algebra, 16 (1970) 40-59

Robson, J.C., "Non-Commutative Dedekind Rings", J. Algebra, 9 (1968) 249-65

________, "A note on Dedekind domains", Bull, London Math. Soc., 3 (1971) 42-46

________, "Decompositions of Noetherian rings", Comm. Algebra, 9 (1974) 345-9

________, Also see Chatters, A.W.; and also Eisenbud, D.

Roos, J.E.,"Sur l'anneau maximal de fractions des AW* - algebres et des anneaux de Baer", C.R. Acad. Sc. Paris., 266 (1968) 120-123

Sandomierski, F.L. "Nonsingular rings", Proc. Amer. Math. Soc. 19 (1968) 225-30

_________, Also see Cateforis, V.C.

Shores, T.S., and Wiegand, R., "Rings whose finitely generated modules are direct sums of cyclics", J. Algebra, 32 (1974) 157-72

Silver, L.,"Noncommutative localizations and application", J. ALgebra, 7 (1967) 44-76

Singh, S., "Modules over hereditary Noetherian prime rings", Canada J. Math. 27 (1975) 867-883

Small, L., "Semihereditary rings", Bull. Amer. Math. Soc., 73 (1967) 656-8

Spircu, T. See Popescu, N.

Strenstrom, B., Rings of Quotients, Grundl. der Math. Wiss., 217, Springer-Verlag, Berlin, Heidelberg, and New York 1975

Storrer, H.H., "Epimorphic extensions of non-commutative rings", Commentarii Math. Helv. 48 (1973) 72-86

_________, "A characterization of Prufer domains", Canada. Bull. Math., 12 (1969) 809-812

Swan, R., Algebraic K-Theory, Springer-Verlag, New York 1968

_________, "Projective modules over rings and maximal orders", Ann. of Math., 75 (1962) 55-61

Tachikawa, H., "A generalization of quasifrobenius rings", Proc. Amer. Math. Soc., 20 (1969) 471-76

Tominaga, H., "Some remarks on π - regular rings of bounded index" Math. J. Okayama Univ., 4 (1955) 135-144

Utumi, "A note on an inequality of Levitzki", Proc. Japan Acad., 33 (1957) 249-251

_________, "Rings whose one-sided quotient rings are two-sided", Proc. Amer. Math. Soc., 14 (1963) 141-6

Walker, E.A. See Faith, C.

_________, "On continuous Rings and Self-injective rings", Trans. Amer. Math. Soc., 118, (1965) 158-173

_________, "Self-injective rings", J. Algebra, 6 (1967) 56-64

Vamos, P., "The decomposition of finitely generated modules and fractionally self-injective rings", J. London Math. Soc., (2), 16 (1977) 209-220

__________, "Sheaf-theoretical Methods in the Solution of Kaplansky's problem", in Applications of Sheaves Fourman, Mulvey, and Scott, eds.), Lecture Notes in Math. Vol. 753, SPringer Verlag, 1979

Warfield, R.B., Jr., "Decomposition of fintiely presented modules", Proc. Amer. Math. Soc., 25 (1970) 167-72

__________, "Serial rings and finitely presented modules", J. Algebra, 37 (1975) 187-222

Webber, D.B., "Ideals and modules of simple Noetherian hereditary rings". J. Algebra, 16 (1970) 230-42

Wiegand R., and Wiegand, S., "Commutative rings over which finitely generated modules are direct sums of cyclics", preprint, U. of Nebraska, Lincoln, Neb. 68588, 1977

Wilkerson, R.W., "Finite dimensional group rings"., Proc. Amer. Math. Soc., 41 (1973) 10-16

Zaks, A., "Some rings are hereditary", Israel J. Math., 10 (1971) 442-450

ABBREVIATIONS AND SYMBOLS

${}^{\perp}X(X^{\perp})$	left (right) annihilators
ACC = a.c.c.	ascending chain condition
$acc^{\perp}$	ascending chain condition on annihilators, p. 1.17
acc ⊕	ascending chain condition on direct sums, p. 1.17
AMVR	almost maximal valuation ring
ann(M)	annihilator of M
ARMC	Algebra: Rings, Modules and Categories I
ART	Algebra II: Ring Theory
Br(k)	Brauer group
CFPF	completely FPF p. 1.2
CFP^2F	completely FP^2F p. 1.2
CQF	completely QF p. 1.11
D(M)	width of M p. 3.10
dim(M)	Goldie dimension
d.c.c.	descending chain condition
E(M)	injective envelope = hull
f.a.e.	following are equivalent
f.g	finitely generated
f.p	finitely presented
FGC	see page I.3
FPF	finitely pseudo Frobenius; see page 1.2
FSI	Fractionally selfinjective
$G_r(R)$	big genus; p. 1.23
$g_r(R)$	little genus; p. 1.23
$\gamma(M)$	genus of M; p. 1.23
HNP	hereditary Noetherian prime

I^{-1}	p. 4.3
$I_f, I_\infty, II_f, II_\infty$, III	p. 3.14
$I <_r R$	rational extension
$t(M)$	torsion submodule; see page 1.19
M	injective hull
M^*	dual module of M
$M^{(I)}$	direct summ of I copies of M
M^I	direct product of I copies of M
$M \approx B$	M is isomorphic to B
mod-A	the category of all right A-modules for a ring A, see page p.1
A-mod	the category of all left A-modules for a ring; A, see page p.1
mod-A ≈ mod-B	category equivalence also denoted Morita equivalence, abbr. M.E. see page 1.2
n.s.	nonsingular
$\nu(M)$	minimal number of generators of M
prindec	principal indecomposable
PID	principal ideal domain
PIR	principal ideal ring
PF	pseudo Frobenius
Q	the field of rational numbers
$Q_c^\ell(R) = Q_{c\ell}^\ell(R)$	left classical ring of quotients of R
$Q_c^r(R) = Q_{c\ell}^r(R)$	right classical ring of quotients of R
$Q^\ell(R) = Q_{max}^\ell(R) = Q_m^\ell(R)$	left maximal ring of quotients
$Q^r(R) = Q_{max}^r(R) = Q_m^r(R)$	right maximal ring of quotients
QF	quasi-Frobenius
RLM	restricted left minimum condition
RRM	restricted right minimum condition
$R = (B,E)$	split-null extension, p. 5.15
$R_1 \overset{\ell}{\sim} R_2$	left equivalent orders, p. 4.3

$R_1 \overset{r}{\approx} R_2$	right equivalent orders, p. 4.3
$R_1 \overset{Q}{\approx} R_2$	equivalent orders, p. 4.3
sing(M)	singular submodule, also denoted Z(M).
s.h.	semihereditary
TF	Torsion free; see page 4.7
VD	valuation domain
VR	valuation ring
UME	Unimodular element
w.r.t.	with respect to
$\mathbb{Z}$	ring of rational integers
$Z_\ell(R)$	left singular ideal of R; also sing(R)
$Z_r(R)$	right singular ideal of R
Z(M)	singular submodule of M
z.d.s.	zero divisors

INDEX

Printed in the United States
By Bookmasters